WIRELESS PERSONAL COMMUNICATIONS
Research Developments

THE KLUWER INTERNATIONAL SERIES IN ENGINEERING AND COMPUTER SCIENCE

COMMUNICATIONS AND INFORMATION THEORY
Consulting Editor
Robert Gallager

Other books in the series:

PLANNING AND ARCHITECTURAL DESIGN OF INTEGRATED SERVICES DIGITAL NETWORKS, A. Nejat Ince, Dag Wilhelmsen and Bülent Sankur
ISBN: 0-7923-9554-9
WIRELESS INFRARED COMMUNICATIONS, John R. Barry
ISBN: 0-7923-9476-3
COMMUNICATIONS AND CRYPTOGRAPHY: *Two sides of One Tapestry*, Richard E. Blahut, Daniel J. Costello, Jr., Ueli Maurer and Thomas Mittelholzer
ISBN: 0-7923-9469-0
WIRELESS AND MOBILE COMMUNICATIONS, Jack M. Holtzman and David J. Goodman
ISBN: 0-7923-9464-X
INTRODUCTION TO CONVOLUTIONAL CODES WITH APPLICATIONS, Ajay Dholakia
ISBN: 0-7923-9467-4
CODED-MODULATION TECHNIQUES FOR FADING CHANNELS, S. Hamidreza Jamali, and Tho Le-Ngoc
ISBN: 0-7923-9421-6
WIRELESS PERSONAL COMMUNICATIONS: *Trends and Challenges*, Theodore S. Rappaport, Brian D. Woerner, Jeffrey H. Reed
ISBN: 0-7923-9430-5
ELLIPTIC CURVE PUBLIC KEY CRYPTOSYSTEMS, Alfred Menezes
ISBN: 0-7923-9368-6
SATELLITE COMMUNICATIONS: Mobile and Fixed Services, Michael Miller, Branka Vucetic and Les Berry
ISBN: 0-7923-9333-3
WIRELESS COMMUNICATIONS: Future Directions, Jack M. Holtzman and David J. Goodman
ISBN: 0-7923-9316-3
DISCRETE-TIME MODELS FOR COMMUNICATION SYSTEMS INCLUDING ATM, Herwig Bruneel and Byung G. Kim
ISBN: 0-7923-9292-2
APPLICATIONS OF FINITE FIELDS, Alfred J. Menezes, Ian F. Blake, XuHong Gao, Ronald C. Mullin, Scott A. Vanstone, Tomik Yaghoobian
ISBN: 0-7923-9282-5
WIRELESS PERSONAL COMMUNICATIONS, Martin J. Feuerstein, Theodore S. Rappaport
ISBN: 0-7923-9280-9
SEQUENCE DETECTION FOR HIGH-DENSITY STORAGE CHANNEL, Jaekyun Moon, L. Richard Carley
ISBN: 0-7923-9264-7
DIGITAL SATELLITE COMMUNICATIONS SYSTEMS AND TECHNOLOGIES: Military and Civil Applications, A. Nejat Ince
ISBN: 0-7923-9254-X
IMAGE AND TEXT COMPRESSION, James A. Storer
ISBN: 0-7923-9243-4
VECTOR QUANTIZATION AND SIGNAL COMPRESSION, Allen Gersho, Robert M. Gray
ISBN: 0-7923-9181-0
THIRD GENERATION WIRELESS INFORMATION NETWORKS, Sanjiv Nanda, David J. Goodman
ISBN: 0-7923-9128-3
SOURCE AND CHANNEL CODING: An Algorithmic Approach, John B. Anderson, Seshadri Mohan
ISBN: 0-7923-9210-8
ADVANCES IN SPEECH CODING, Bishnu Atal, Vladimir Cuperman, Allen Gersho
ISBN: 0-7923-9091-1
SWITCHING AND TRAFFIC THEORY FOR INTEGRATED BROADBAND NETWORKS, Joseph Y. Hui
ISBN: 0-7923-9061-X

WIRELESS PERSONAL COMMUNICATIONS
Research Developments

edited by

Brian D. Woerner
Theodore S. Rappaport
Jeffrey H. Reed
Virginia Polytechnic Institute
& State University

KLUWER ACADEMIC PUBLISHERS
Boston / Dordrecht / London

Distributors for North America:
Kluwer Academic Publishers
101 Philip Drive
Assinippi Park
Norwell, Massachusetts 02061 USA

Distributors for all other countries:
Kluwer Academic Publishers Group
Distribution Centre
Post Office Box 322
3300 AH Dordrecht, THE NETHERLANDS

Library of Congress Cataloging-in-Publication Data

A C.I.P. Catalogue record for this book is available
from the Library of Congress.

Printed on acid-free paper.

Printed in the United States of America

TABLE OF CONTENTS

PREFACE   ix

I   TUTORIALS IN WIRELESS COMMUNICATION

        1.   Radio Wave Propagation for Emerging Wireless
                Personal Communication Systems
                T. S. Rappaport and S. Sandhu   1

        2.   A Survey of Adaptive Single Channel
                Interference Rejection Techniques for
                Wireless Communications
                J. D. Laster and J. H. Reed   29

        3.   Spread Spectrum for Wireless Communications
                R. M. Buehrer and B. D. Woerner   55

II   DIGITAL SIGNAL PROCESSING AND SPREAD SPECTRUM
        FOR WIRELESS COMMUNICATIONS

        4.   CMA Adaptive Array Antenna Using Transversal
                Filters for Spatial and Temporal Adaptibility in
                Mobile Communications
                N. Kikuma, K. Hachitori, F. Saito and N. Inagaki   75

        5.   The Effect of Adjacent Cell Interference, Power
                Control and Error Correction on a CDMA
                Cellular System
                A. Sathyendran, K. W. Sowerby and M. Shafi   85

        6.   The Design and Development of a Spread Spectrum
                Digital Cordless Telephone for the Consumer Market
                P. S. McIntosh   97

7. A Principled Framework for Narrowband Mobile
    Digital Communications
    M. P. Fitz and J. P. Seymour   107

8. BER Performance of Adaptive RAKE Receiver
    Using Tap Weights Obtained by POCS
    Deconvolution Technique
    S. Hussain, Z. Kostic and B. Gopinath   119

III PROPAGATION

9. A Macrocell Model Based on the Parabolic
    Diffusion Differential Equation
    Jan-Erik Berg   131

10. A Nano-Cellular Local Area Network Using
    Near-Field RF Coupling
    A. Demers, S. Elrod, C. Kantarjiev and E. Richley   141

11. Radiowave Propagation Measurements for
    Sharing Spectrum Between Point-to-Point
    Microwave Radios and Personal
    Communications Systems
    S. Y. Seidel, D. M. J. Devasirvatham, R. R. Murray,
    H. W. Arnold and L. G. Sutliff   157

12. Estimating Regions of Service in Wireless Indoor
    Communication Systems
    P. M. Cartwright and K. W. Sowerby   167

13. Adaptive Propagation Prediction Using Lee's
    Model in a Non-Homogeneous Environment
    R. Lopez and K. Vlahodimitropulos   179

14. An Interactive System for Visualizing Wireless
    Communication Coverage Within Buildings
    M. A. Panjwani, A. L. Abbot and T. S. Rappaport   185

IV   TECHNICAL EVALUATION OF EMERGING MOBILE
WIRELESS SYSTEMS

15.   Influence of Slow Frequency Hopping and Antenna
Diversity Techniques in European GSM System
F. de Seze   197

16.   Network Architecture and Radio Link
Performance of MOBITEX® Systems
R. Alavi and M. M. Khan   207

17.   Performance Analysis of Digital FM in
Simulcast Environment
R. Petrovic, W. Roehr, D. Cameron and K. Tano   215

18.   Improving Throughput and Availability of
Cellular Digital Packet Data (CDPD)
J. M. Jacobsmeyer   227

V   ISSUES FOR DEFENSE CONVERSION INTO THE
WIRELESS INDUSTRY

Introduction - The Impact of Defense Conversion on the Wireless
Industry: *Corporate and Government Perspectives*
Jeffrey H. Reed   239

19.   ARPA's Prospective of Defense Conversion
Opportunities for the Wireless Industry
Randy Katz   241

20.   ESL's Electronic Warfare Capability for the
Cellular Industry
Bruce Cole   247

21.   Hazeltine Corporation Defense Conversion to
Commercial Wireless Communication Products
Henry L. Bachman   251

22.   Melting Swords Into Plowshares
Jerome C. Woodward   255

23.   Challenges in Defense Conversion
       Joe Kennedy   **263**

VI   COMPETITIVE ISSUES FOR PCS

Introduction - Competitive Issues in PCS
George Morgan   **267**

24.   Business and Regulatory Issues in PCS
       Byron Marchant   **271**

25.   ESMR, SMR and the Transition to PCS
       Corey Eng   **279**

26.   The Importance of Unlicensed Spectrum
       Jim Lovett   **285**

INDEX   **293**

PREFACE

The area of personal and wireless communications is a burgeoning field. Technology advances and new frequency allocations for personal communication services (PCS) are creating numerous business and technical opportunities. At the Mobile and Portable Radio Research Group (MPRG) we see these opportunities from many different perspectives, as educators of wireless engineers, researchers, and as advisors to numerous wireless companies. It is becoming clear that an essential requirement for exploiting opportunities is the ability to track the dramatic changes in wireless technology. It is for the dissemination of late breaking research that we host an annual symposium on wireless personal communications each year in June on the campus of Virginia Tech.

The symposium offers an informal atmosphere where practicing engineers, students and researchers can gather and discuss trends and opportunities. This year, the symposium attracted nearly 200 participants from many different parts of the world. This book is a summary of the ideas presented at the Fourth Virginia Tech Symposium on Wireless and Personal Communications.

This book is composed of six parts. Part I contains several tutorial chapters on key areas in wireless communications written by the MPRG faculty and students. The first of these tutorials is on radio wave propagation for emerging wireless personal communication systems and is written by Ted Rappaport and graduate student Sandip Sandhu. This tutorial conveys a wealth of experience in measuring propagation losses for various outdoor and indoor environments, and points toward research challenges that face propagation engineers as PCS is deployed. The second tutorial, written by Jeff Reed and PH.D. student Jeff Laster, is a comprehensive study of emerging DSP-based interference rejection techniques for single channel (antenna) systems. Interference is the primary capacity limiting factor for cellular and personal communication systems. The third tutorial is on spread spectrum wireless communications and is written by Brian Woerner and doctoral student Mike Buehrer. This tutorial explains the concept of spread spectrum, modeling techniques for spread spectrum, and current applications and research issues for spread spectrum systems.

Part II focuses on digital signal processing and spread spectrum, two means of creating interference and multipath robust communications. The first chapter in this part written by Nobuyoshi Kikuma, Kazuya Hachitori, Fuminobu Saito, and Naoki Inagaki, presents the performance of a constant modulus algorithm adaptive array for interference rejection. The next chapter by Sowerby, Sathyendran, and Shati examines the impact of adjacent cell interference, power control, and error correction of the performance of CDMA cellular radio. Stuckey McIntosh's chapter describes the design of a direct sequence spread spectrum cordless telephone that operates in the ISM band. Michael Fritz and James Seymor of Purdue University describe optimal demodulation techniques for narrowband communications in the presence of fading. Finally Shehzad Hussain, Zoran Kostic, and B. Gopinath use projection onto convex sets as a method of deconvolution to obtain a superior RAKE receiver for multipath corrupted signals.

Part III concerns propagation aspects of wireless communications. Jan-Erik Berg, from Ericsson Radio Systems, presents a parabolic diffusion differential equation for estimating path loss in macrocell systems. Alan Demers, Scott Elrod, Christopher Kantarjiev and Edward Richley, discuss a nano-cellular local area network using near-field RF coupling at an operating frequency of 5.3 MHz. This technique is simple and inexpensive and offers relatively good data rates, excellent frequency reuse, and the possibility of user location. Scott Seidel, D. M. J. Devasirvatham, R. R. Murray, H. W. Arnold and L. G. Sutliff, examine the propagation characteristics in New York City to ascertain if PCS systems can co-exist with point to point microwave systems. The results in this Bellcore sponsored research project are not very encouraging for the feasibility of PCS and microwave spectrum sharing. Paul Cartwright and Kevin Sowerby discuss the effects of obstacles, noisy office equipment, and signal variability on indoor wireless systems. Robert Lopez and Konstantinos Vlahodimitropulos discuss a method for improving the accuracy of signal propagation prediction in an environment of variable building/clutter density based on Lee's model. Manish Panjwani and other Virginia Tech researchers describe an interactive tool for assessing the RF coverage of a multifloor building using an AutoCAD description of the building.

Part IV discusses the performance of emerging wireless systems. GSM is a world-wide cellular standard, yet its full capability is yet to be demonstrated. Fabrice de Seze from Alcatel Radiotelephone in France describes a simulation tool for evaluating the performance of GSM with slow frequency hopping and antenna diversity and provides some results obtained using this simulation tool. Wireless data communications is one of the fastest growing segments of the wireless market. Reza Alavi and Mobeen Khan of RAM Mobile Data provide an overview of the MOBITEX wide area packet-switched data communication system. Rade Petrovic, Walter Roehr, Dennis Cameron and Kunio Tano, review the effects of velocity, transmitter frequency offset, differential propagation delay and other parameters of FM simulcast systems. Cellular Digital Packet Data (CDPD) is an important U. S. standard for cellular data communications. Jay Jacobsmeyer describes a technique for varying the data to compensate for multipath fading effects and to increase the service area for CDPD systems.

Part V describes the opportunities and pitfalls of defense conversion from the perspective of several U. S. Defense firms that have successfully made the transition to commercial wireless. The perspective of the Advanced Research Projects Agency (ARPA), a federal government organization with a mission to foster this conversion, is also presented. The chapters are based on a panel session composed of representatives from defense industries that have successfully made this transition including Henry Bachman from Hazeltine, Bruce Cole from ESL, Joe Kennedy from ERA, and Jerry Woodard from Signal Science. Finally, Randy Katz from UC Berkeley and ARPA discusses defense conversion from the perspective of a proposal reviewer within the Defense Technology Reinvestment Program (TRP), a multi-billion dollar program initiated by the U. S. government to facilitate defense conversion. At the MPRG, we have seen significant interest in defense conversion as a business strategy. The advice provided by the panelists is quite insightful to those involved in the defense conversion process.

Part VI is a summary of the panel session on the competitive issues of PCS. This part includes a discussion on business and regulatory issues in PCS by Byron Marchant of the FCC. Corey Eng of MCI provides an overview of the impact of ESMR and SMR on PCS. Jim Lovette from Apple Computer makes the case for unlicensed spectrum for emerging data communications applications.

We would like to thank the many people whose work has made our annual symposium and this book possible. The quality of research presented in this book is due to the diligent efforts of the individual authors and the MPRG Industrial Affiliate companies: AirTouch Communications, Apple Computer, AT&T Bell Laboratories, Bell Northern Research, Bellcore, BellSouth Wireless, FBI, Grayson Electonics, GTE, MCI, Motorola, National Semiconductor, Rockwell International, Southwestern Bell, and US West, who continue to support our mission of education and research. Jenny Frank has once again successfully coordinated the 1994 symposium and assembled the chapters which appear in this book. Annie Wade's tireless efforts in transcribing panel sessions for inclusion into this book have made a significant contribution to the quality of the manuscript. The symposium would not have been possible without the dedicated efforts of Prab Koushik, Kathy Wolfe, and Valarie Caple. Lastly, the students of MPRG have provided tremendous support for the Symposium and this book.

1

RADIO WAVE PROPAGATION FOR EMERGING WIRELESS PERSONAL COMMUNICATION SYSTEMS

Theodore S. Rappaport and Sandip Sandhu

Mobile and Portable Radio Research Group
Bradley Department of Electrical Engineering
Virginia Polytechnic Institute and State University
Blacksburg, Virginia 24061
e-mail: N9NB@VTVM1.CC.VT.EDU

Abstract

This survey paper summarizes recent radio propagation measurements and models for emerging wireless personal communication systems. Both indoor and microcell propagation environments are considered.

1 Introduction

The past decade has seen a phenomenal growth in wireless communications. Wireless technology is permeating business and personal communications across the globe and the demand is driving availability and performance to new levels. Consumers are demanding small handheld or pocket communicators to meet their wireless voice and data communications needs. Cellular radio systems, paging systems, mobile satellite systems, cordless telephones and future personal communications systems (PCS) all aim to provide ubiquitous access. Even with numerous standards for each of these wireless systems, the companies involved in providing such services have shown annual growth rates of 30-50%[1]. The demand for ubiquitous communications has led to the development of new wireless systems like PCS (Personal Communication Systems), W-LAN (Wireless Local Area Networks), W-PBX (Wireless Private Branch Exchanges) and parasitic cellular systems. Consumers have also indicated a strong interest in small handheld devices like the PDA (personal digital assistant) and PICD (personal information and communication device) which provide the functionality of calendar, clock, fax, notepad, pager and cellular phone all in a small, easy to use package. These devices are being developed with expected delivery in 1995. With a projected cellular radio penetration of 100 million users in the U.S. alone by the year 2000 [2], service providers must offer small wireless devices that are inexpensive and easy to use.

In cellular radio systems, the coverage region is divided into small areas called cells (typically with radius of 2 to 10 km). Each cell has its own base station and specific set of channels [3]. These channels are used to provide duplex communications. To increase spectrum utilization, the cellular systems use the frequency reuse concept in which one frequency (or a group of frequencies) can be used again in a different cell. Two such cells that use the same group of frequencies are called co-channel cells, and must be separated by a minimum distance to keep co-channel interference below acceptable limits. As 800 MHz cellular systems mature, their capacity can be increased to accommodate more users by cell splitting, where each cell is split into smaller cells with lower transmitter power and antennas with downtilted radiation patterns. This technique is already being used in densely populated areas, where cell sizes are sufficiently small to be considered microcells [3].

In early 1995, the FCC will auction 160 MHz of spectrum (120 MHz for licenced use and 40 MHz for unlicensed use) for PCS in the 1.8 to 2.2 GHz band [4]. Three 1 MHz PCS bands at 901-902, 930-931 and 940-941 MHz will also be made available for narrowband PCS. Currently in the U.S, the ISM bands of 902-928, 2400-2483 and 5700-5900 MHz can be used to provide a wide range of voice and data communication services. Consequently, a large number of different systems providing different services are now available in the market. A few different products in the unlicensed ISM band use the same set of frequencies. This has lead to a set of new propagation issues that must be addressed to be able to design and install systems and to be able to evaluate their performance in a noisy environment.

PCS differs from a cellular system in that it will provide a range of telecommunication services by using low power base transmitters (10 - 100 mW) and a smaller cell size (200 - 400 m) in the 1.8-2.2 GHz band[4]. PCS will provide two way calling plus high bandwidth data, voice and video transmission services to a large number of users in a small area. PCS will merge cellular and cordless telephone concepts and will likely provide data and video transmission services.

W-LAN will allow networking of fixed and portable computers via a wireless data link inside office buildings, airports, banks and other locations where flexible, reconfigurable computer networks are needed on demand. W-LAN products already exist for operation in the U.S ISM band for data rates up to 2Mbps. Licenced systems in the 18 GHz band also provide W-LAN capabilities at data rates close to the 10 Mbps Ethernet standard, with complete Ethernet handshaking. A portion of the U.S. PCS band will be dedicated to W-LAN services for laptop computers, and Europe's new HIPERLAN initiative in the 5.6 GHz band will support 20 Mbps data rates for W-LAN applications.

W-PBX will provide ubiquitous access to telephone services inside a building so that anyone may use a portable telephone in the work place. Already, numerous W-PBX trials are being carried out by the regional Bell operating companies using products in the ISM, cellular, and PCS bands, and customer satisfaction is extremely high. Particularly in hospitals and factories, where employees are constantly on the move, the ability to send and receive messages and telephone calls is valuable. The cellular industry, in order to provide services comparable to PCS, has developed indoor *parasitic systems* to provide telephone coverage inside buildings using conventional cellular handsets and cellular spectrum. When inside a building, users place and receive calls at no cost, and enjoy the full functionality of a W-PBX. The same handset is used as a conventional cellular telephone once outside the building. The parasitic systems are based on the fact that buildings attenuate RF energy which is radiated from co-channel cellular transmitters located outside of buildings. Therefore, it is possible to have self-controlled indoor cellular systems which use the same frequencies as the cellular system outside of buildings. The parasitic systems thus borrow unused cellular channels for increased cellular capacity inside buildings. Such indoor cellular systems rely on RF isolation from the outdoor cellular systems and use *sniffer receivers*, such as the one described in [45], which scan all of the cellular channels inside a building to determine the available indoor channels. A channel is deemed available if both the forward and reverse frequency pair have low received signal strengths. Clearly, the location of the sniffing receivers will determine the specific number of channels deemed available within a building at any given time. For such parasitic systems, the indoor capacity and RF isolation from the outdoor cellular system will determine its reliability and performance, and may be subject to change as the outdoor cellular system matures.

To implement any of the personal wireless systems described above, there is demand for RF channel models based on electromagnetic theory and measurements be developed to determine propagation characteristics for an arbitrary installation. It is important to predict propagation power loss, interference, spatial distribution of power, and RF dispersion in time and frequency domains in order to properly anticipate conditions in static channels, mobile channels (for walking speeds) and for both narrowband (flat fading) and wideband (frequency selective) channels. Recently, a number of measurements have been carried out to determine the statistical and site specific properties of RF propagation for personal communications in order to develop and verify propagation models.

Design of parasitic systems and performance evaluation of emerging wireless systems require that building penetration of radio waves be modeled. Accurate prediction of received signal strength inside a building due to transmitter outside the building, and vice-versa, will also help in understanding the co-system problems that arise when two or more systems are installed in two adjacent buildings. Such a prediction will also allow optimum frequency reuse and an efficient utilization of limited bandwidth resources.

Although typical PCS channel characteristics are known and a few statistical indoor propagation models are available, more accurate prediction techniques are necessary to provide accurate capacity and coverage predictions in a given situation. As propagation models evolve, site specific channel impulse response prediction methods may be used instead of time consuming measurements for accurate RF predictions. A few site specific propagation models for indoor [7] and microcellular environments [8] have shown promising results. Ray tracing seems to be form the basis for most of the site specific predictions.

This survey paper provides an overview of some recent propagation results for emerging wireless systems. Section 2 discusses the propagation issues relating to communication systems in and around buildings. Section 3 summarizes propagation characteristics in and around buildings. Section 4 discusses microcell propagation models. Section 5 deals with building penetration issues and related phenomena. Section 6 concludes the paper.

2 Propagation Issues for Personal Communication Systems

Wireless systems tend to be interference limited, rather than noise limited. That is, while link budgets are important for determining coverage areas and power levels between specific transmitters and receivers, it is the interference from co-channel transmitters that ultimately limits the capacity and performance of wireless systems.

There are four basic propagation scenarios for PCS, as shown in Figures 1 to 4. These may be classified as indoor/indoor, indoor/outdoor, outdoor/indoor and outdoor/outdoor. The **indoor/ indoor** propagation problems occur when two or more indoor wireless systems are operating in the same building in the same frequency band. These co-channel systems need to be able to co-exist, even if they employ different multiple access methods. Another example where indoor/ indoor propagation issues become important is for the case where co-channel transmitters are used on different floors or in different rooms of the same building. The level of adjacent channel, co-channel and co-system interference within buildings needs to be understood. Analysis of propagation between different floors in buildings will help understand the co-channel interference problem due to systems using same spectrum but operating on different floors.

Outdoor/indoor issues arise in systems like the parasitic cellular system. As the outdoor cellular systems mature and the cell site transmitters come closer to the building (due to cell splitting), the parasitic system performance may degrade, since the number of interference-free channels available for use inside the building may decrease over time. For an accurate forecast of future capacity, a proper model for the building penetration of RF waves must be developed. For parasitic cellular systems, performance degradation may also be experienced by the outdoor system due to the radiation of the indoor system in a multi-floor office building. This will give rise to the **indoor/outdoor** propagation problems, where sources inside a building may interfere with co-channel systems operating outdoors.

Outdoor/outdoor propagation issues arise for two or more systems that operate in adjacent buildings, in streets, or on a campus, while using the same frequency band. PCS and microcellular systems must be concerned with this scenario, as must indoor W-LAN systems in tall buildings.

3 Received Signal Characteristics In and Around Buildings

Radio signals in and around buildings are attenuated as they pass through walls, buildings and objects, and since energy may arrive from different directions and at different times due to reflections from surrounding objects, there is multipath propagation. A time-varying channel occurs due to receiver mobility or changes in the channel caused by moving vehicles, people, or other objects in motion. These factors cause deep fades in the received signal which varies with time, and creates intersymbol interference due to pulse spreading [9].

Figure 1. Indoor/indoor co-system interference

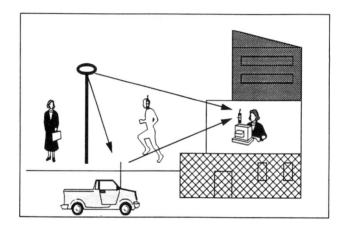

Figure 2. The Outdoor/indoor interference

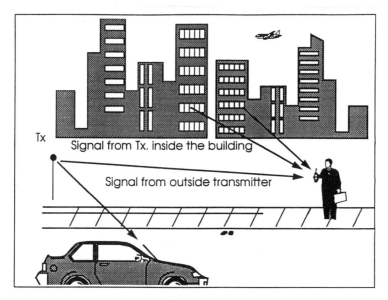

Figure 3. Indoor/outdoor interference

Figure 4. Outdoor/outdoor interference

3a) Small scale fading inside buildings

Within a building, a direct path may exist between the transmitter and receiver. Signals may also arrive at the receiver antenna after reflection from surrounding or nearby objects. The vector addition of all multipath components at the receiver antenna may increase or decrease the resulting signal amplitude. For a mobile channel, the time varying propagation distance changes the phase relationships of individual multipath components, resulting in deep fades in the received signal envelope as shown in Figure 5. It has been shown that the cumulative distribution function (CDF) of the received signal envelope is described by a Rayleigh distribution if no direct line of sight (LOS) path exists between the receiver and transmitter [11]. If a direct path exists between the receiver and the transmitter, the statistics of the signal envelope is Rician [9] over a small measurement area. When Rayleigh fading over small scale distances is shown to exist, it occurs for all antenna polarizations [3].

Reference [18] reported experiments to determine statistics related to the random variation of a CW signal received at a fixed location in an indoor environment. The measurements were performed in two different types of buildings. It was originally anticipated that CDFs of the received signal for fixed receiver locations with no LOS path to transmitter should correspond well with the Rayleigh distribution. Instead, distribution functions computed from the data at all locations were found to be Rician with different values of K [13],[18]. Specifically, the small scale fading of the envelope was found to be Rician, with a K ratio of about -2 dB. The 10 and 90 percentile levels of this distribution lie within -7 dB and 4 dB of the median. Even less small scale fading was observed for fixed terminals in factory buildings [13].

3b) Path loss inside buildings

When the received signal at a specific location is averaged over a local measurement track having a length of between 5 and 20 wavelengths, a local average signal level can be measured and used to determine mean signal strengths. When a large number of local average signal levels are plotted (in dB) against the distance between the transmitter and receiver (T-R separation), a general distance-dependent model can be applied to describe the RF path loss. Path loss describes the attenuation in the propagation environment, and is a measure of the channel attenuation at a specific location relative to a close-in free space reference [1]

[11]. Measurements have shown that a good path loss model at a specific T-R separation is given by

$$\overline{PL}\,(d)\,[\text{dB}] \;=\; PL\!\left(d_0\right) + 10 \times n \times \log_{10}\!\left(\frac{d}{d_0}\right) + X_{\sigma}\;[\text{dB}] \tag{1}$$

where $\overline{PL}\,(d)$ is the mean path loss in dB at a T-R separation of d meters, $PL(d_0)$ is the free space path loss at a close-in reference distance d_0, and n is a path loss exponent which describes how fast the mean path loss increases with distance [1],[7]. For indoor measurements, d_0 should be in the far field of the antenna, and $PL(d_0)$ is due to free space propagation from the transmitter to d_0. The term X_{σ} in equation (1) describes the variation of path loss at a specific distance for independent measurement locations made throughout a large area. This variation of the large-scale path loss is often log-normally distributed about the mean path loss [9], where X_{σ} is a zero-mean log-normally distributed (normal in dB) random variable with standard deviation σ in dB. Linear regression is normally used to compute values of the path loss exponent n and the standard deviation σ (in dB) about the best-fit mean path loss model in a minimum mean square error sense (MMSE) for the measured data. Figure 6 shows a typical scatter plot obtained from measured data [7] and the best fit mean path loss exponent in a MMSE sense.

Indoor channel measurements indicate that large signal variations occur in different buildings. Table 1 shows path loss exponent variation from 1.4 to 3.8. The variation may be attributed to the type of building. It may be observed that open plan buildings, like factories, where the reflections from nearby metal objects are greater, have a smaller path loss exponent in comparison to the value for an office building where one expects more obstructions and partitions, and hence greater attenuation.

In [18] it was observed that in LOS configurations in office buildings, the path loss varies as $d^{1.8}$. Average diffraction losses of 10-15 dB were observed in cases where line of sight was blocked by one or more walls. A dynamic range of 30 dB was found for fading inside the building and the temporal variations were slow (4Hz). Other measurements have reported a value for the mean path loss exponent, n, that is different when a line of sight path exists (LOS) as compared to the obstructed case (OBS case) where no line of sight path exists. Reference [7] reports the following observations for 914 MHz inside various office buildings.

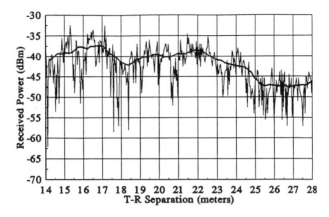

Figure 5. A typical received waveform inside buildings showing small scale fading with deep fades (20 dB and more) and average signal strength over local area

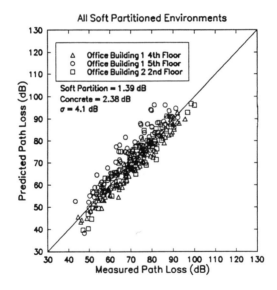

Figure 6. Scatter plot of measured and predicted path loss [7] at 914 MHz

TABLE 1. Path loss measured in different buildings

Building	n	σ (dB)	Frequency (MHz)	Reference
Grocery store	1.8	5.2	914	12
Retail store	2.2	8.7	914	12
Open-plan Factories	2.2	7.9	1300	13
Open-plan Factories	1.4-3.3		4000	14
Open-plan Factories B	2.0	3.7	1300	15
Open-plan Factories B	2.1	4.0	1300	15
Open-plan Factories C	2.4	9.2	1300	15
Open-plan Factories C	2.1	9.7	1300	15
Suburban office building open plan	2.4	9.6	915	16
Suburban office building open plan	2.6	14.1	1900	16
Suburban office building soft partition	2.8	14.2	915	16
Suburban office building soft partition	3.8	12.7	1900	16
Office building hard partition	3		850	17

TABLE 2. Different path loss values for LOS
and OBS case at 914 MHz [3]

Topography	n	σ (dB)
Open-plan LOS	1.9	6.5
Open-Plan OBS	2.4	5.5
Hard-Partioned LOS	2.3	6.2
Hard Partioned LOS	3.4	7.3
Soft-Partioned LOS	2.6	6.7
Soft-Partioned OBS	3.2	7.4

Since it is evident that objects and partitions inside a building affect the path loss exponent, a path loss model depending on site specific factors should give accurate results. Work
in [12] presented a quantitative model that predicts the effects of walls, office partitions,
floors and building layout on path loss at 914 MHz. In [12], thousands of signal strength measurements from four base station locations in a multi-story building were used to develop a
simple site specific model which adds an attenuation factor based on the number of partitions
between the transmitter and the receiver, and assumes free space propagation in unobstructed

portions of the building. The standard deviation between the measured and the predicted path loss was found to be 5.8 dB for the entire data set. Average floor attenuation factors (FAF) were found to be 12.9 and 16.2 dB for one floor between transmitter and receiver in two different buildings. The experimental values for the floor attenuation factors are shown in Table 3, and are taken from [12]. For measurements on the same floor, attenuation factors (AF) were found to be 1.4 dB for each cloth covered office partition and 2.4 dB for each concrete wall between the transmitter and receiver. An indoor design tool based on building blue prints and the models in [12] is described in [42].

The floor attenuation factor can also be calculated from theory developed in [37], which explains propagation between a transmitter and receiver located on different floors of a building. It was shown that propagation in such a case can occur via two propagation paths: the transmission path through floors; and, diffracted paths along the outside walls of the building. The path loss was found to be different for the two propagation mechanisms. Measurements also indicated that little power penetrated floors with steel panels, and significant power was carried by the diffracted paths along the building exterior. Path loss increased with the number of floors that separated the transmitter and receiver.

TABLE 3.   Average Floor Attenuation Factor (FAF) [12]

Location	FAF (dB)	σ (dB)
Office building 1		
Through 1 floor	12.9	7.0
Through 2 floors	18.7	2.8
Through 3 floors	24.4	1.7
Through 4 floors	27.0	1.5
Office building 2		
Through 1 floor	16.2	2.9
Through 2 floors	27.5	5.4
Through 3 floors	31.6	7.2

3c) Time dispersion in indoor channels

Power delay profiles are measurements of received power as a function of propagation delay in a multipath channel, and are averaged over time or space to determine multipath characteristics. Three parameters are often used to describe the temporal spread of the channel. The mean excess delay is the first moment of the power delay profile, and describes the average propagation delay relative to the first arriving signal component. The rms delay spread measures the temporal spread of the power delay profile about the mean excess delay. The excess-delay spread (X dB) indicates the maximum delay, relative to the first arriving signal component, at which the multipath energy falls to X dB below the peak received level [19]. These parameters are loosely related to outage and bit-error rates for different digital modulation schemes that do not use equalization [1]. A rule of thumb is that a bit error rate (BER) of less than 10^{-3} will occur if the channel rms delay spread is less than 0.2 of the symbol duration. Figure 7 illustrates a typical power delay profile for PCS channels, and shows that only a few significant multipath components may exist. However, in general, a large number of multipath components will exist, and accurate simulations of digital wireless communication systems rely on knowing the channel time dispersion to resolutions greater than the symbol duration.

Saleh and Valenzuela [5] reported the results of indoor propagation measurements between two vertically polarized omni-directional antennas located on the same floor of a medium-sized office building using 10ns, 1.5 GHz, radar like pulses. The method of measurement involved averaging the square-law-detected pulse response while sweeping the frequency of the transmitted pulse. Thus, the multipath power delay profile was resolvable to within 5 ns.

The results in [5] show that (a) the indoor channel is quasi-static, or very slowly time varying, (b) the statistics of the channel impulse response are independent of the polarization of the transmitter and receiver antennas if there is no line of sight path between them, (c) the maximum observed delay spread in the building is 100-200 ns within rooms, and 300 ns in hallways, (d) the measured rms delay spread within rooms has a median of 25 ns and a maximum of 50 ns, (e) the signal attenuation with no line of sight path varies over a 60 dB range and obeys a path loss law given in equation (1) with an exponent between 3 and 4.

Work in [5] also gives a simple statistical impulse response model for indoor channels which was shown to fit the measurements and may be extended to other buildings. The model assumes that the multipath components arrive in clusters. The received amplitude of each component is an independent Rayleigh random variable with variance that decays exponentially with propagation delay as well as with time delay within a cluster. The corresponding phase angles for each component are independent uniform random variables over [0,2π]. The clusters and multipath components within a cluster form a Poisson arrival process with different rates. The multipath components and the clusters have exponentially distributed inter-arrival times. The formation of the clusters is related to the building structure while the multipath components within each cluster are formed by multiple reflections from objects in the vicinity of transmitter and the receiver.

Work in [13] reported results of measurements at 1300 MHz in five factory buildings. Multipath spreads ranged from 40 to 800 ns. Mean excess delay and rms delay spread values ranged from 30 to 300 ns with median values of 96 ns in LOS paths and 105 ns in obstructed paths. These values were much greater than the ones obtained in [5]. The difference was attributed to the difference in the type of building measured. Delay spreads were found to be uncorrelated with T-R separation but were found to be affected by factory inventory, building construction materials, building age, wall locations and ceiling heights. Measurements in a food processing factory that manufactures dry-goods, with considerably less metal inventory than other factories, had a root mean square delay spread that was half of those observed in factories producing metal products. Newer factories, which incorporate steel beams and steel reinforced concrete in the building structure, have larger delay spread than older factories which use wood and brick for perimeter walls. The measurements showed a worst case rms delay spread of 300 ns inside a modern factory building. This implies that for bit error rates below 10^{-3}, baseband data rates are limited to about 150 kbps for binary rectangular pulse shapes and no equalization [13]. Binary raised cosine pulse shapes can increase the unequalized throughput to about 250 kbps for the same channel. Average factory path loss was found to be a function of distance to the power of 2.2 [13].

The measurements in [13] used 10 ns pulses that were detected by a square law envelope detector to obtain numerous power delay profiles and other parameters. The data suggested that radio propagation in factory buildings may be described by a hybrid geometric/statistical model that accounts for both specular reflections from walls and ceilings and random scattering from inventory and equipment. In [43], a detailed statistical channel impulse response model, which includes spatial correlation and complete descriptions of the time-varying nature of the indoor channel, is presented. The resulting model, called SIRCIM (Simulation of Indoor Radio Channel Impulse response Models) is being used by many researchers.

Table 4 summarizes a few recent results for the measured rms delay spread inside many buildings.

TABLE 4.   Median and maximum rms delay spread in buildings

Environment	Median rms delay spread (ns)	Max. rms delay spread (ns)	reference
Brick office building	26-30	<70	44
Concrete office building	28-29	<70	44
Office buildings	25	50	5
Office buildings	50	218	17
In factories (LOS)	96	300	13
In factories (OBS)	105	300	13

Work in [21] presented a model of the impulse response of indoor channels using measurements in the frequency range 900-1300 MHz and a resolution of 2.5 ns. The results show that the arrival time of the multipath components is characterized by a modified Poisson distribution and that the rms delay spread over large areas is normally distributed with mean values that increase with increasing T-R separation. This is contrary to the results from [5] and [13] as discussed above.

The dependence of rms delay spread on the transmitter and receiver separation was further studied by Bultitude, et. al. [22]. The results showed that the relationship is non-monotonic and has a maximum at a range that depends on building dimensions. Measurements at 950 MHz also showed that the presence of furniture and other reflecting surfaces inside the building can significantly change the rms delay spread at a specific location.

3d) Site-specific propagation prediction inside buildings

A new technique for the prediction of channel responses inside buildings uses blue prints, or more simplified diagrams of the floor plan, in order to provide a deterministic impulse response estimate anywhere in the building. By using the concept of ray tracing, rays may be launched from a transmitter location, and the interaction of the rays with partitions within the building may be modeled using well known reflection and transmission theory.

Either brute force ray tracing methods [46] or ray tracing based on image theory [47], [48] can be used to provide extremely accurate predictions. Rays, representing tubes of energy, can be tested for intersection at any point on the building floor plan. Ultimately, the accuracy of such techniques relies on the accuracy and detail of the site-specific representation of the propagation medium. While this technique requires greater detail than statistical propagation models, it exploits the recent availability of graphical data bases and interactive computing environments, and lends itself extremely well to efficient, parallel processing [46]. It should be noted that the use of ray tracing makes use of the fact that all objects of interest within the propagation environment are larger than a wavelength, thus obviating more exhaustive and impractical methods which solve Maxwell's equations using finite difference techniques.

The above results show that characterization of the indoor radio channel is a complex task. The dynamic temporal variation of the received signal in an indoor environment also creates problems for a reliable link at a given time or location. Fortunately, there are a large number of recent statistical models that can be used for preliminary design and analysis. Site-specific techniques hold great promise for fast and accurate predictions of indoor radio coverage in the future.

4 Path loss in microcells

Radio coverage in urban areas will be improved by using low power microcells with small, well controlled coverage footprints. Microcell base stations will be mounted on lamp posts, street signs, or at other locations where a high user density is expected, and will be connected to the public telephone network via wired (optical fiber), twisted pair or wireless (microwave) links.

The propagation characteristics of microcells are much different than that of conventional, large cell systems. Instead of a simple mean path loss model given in equation (1), a double regression path loss model has been used to predict the path loss with distance [10] [39]. In the double regression model, path loss increases slowly (n is approximately 2) with distance until a breakpoint is reached. After the breakpoint, path loss increases more rapidly with distance (n can range between 3 and 9). The breakpoint depends on antenna height and frequency. This phenomenon is often described by a two ray and four ray propagation model

Measurement Location : Norris 1 F

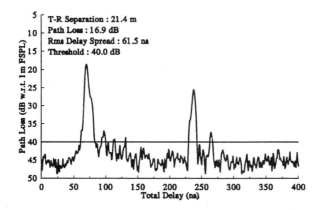

Figure 7. A typical power delay profile and statistical parameters
used to describe measurements [7]

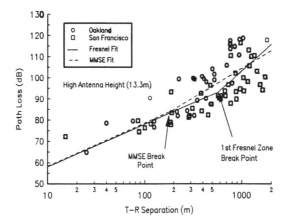

Figure 8. Path loss vs. T-R separation for microcells [10]

[16]. The two ray model considers a direct path and a ground reflected path, and the break-point is determined according to Fresnel zone theory. For the four ray model, two additional wall reflected rays are added, and the breakpoint distance changes according to the street width. This simple model is very appropriate for microcellular systems since the antennas are mounted at heights below the roof line of the surrounding buildings. The geometry of the surroundings make the multi-ray model plausible for computing the regression lines and breakpoint distance.

In [40], 905.15 MHz narrow band measurements were made with omni-directional antennas mounted at 5 meters and 1.5 meters, and breakpoints at 200 and 300 m, respectively, were found. The fading statistics of the CW signals was found to be Rician in microcells, with the Rice K-factor varying over a measurement path. Rician statistics provide a good description for the small scale, narrow band fading in microcellular channels since the dominant LOS component prevents extremely deep fades. Microcellular measurements at 870.15 MHz and 1.8 GHz show that path loss at both frequencies is very similar, except that the spread, or variation, of path loss for 1.8 GHz signals is larger [40]. It is also observed that the 1.8 GHz signal is more sensitive to shadowing than the 870.15 MHz signal. That is, when an obstructing object such as a bus passes between the transmitter and receiver, deeper fades are observed at 1.8 GHz as compared with 870.15 MHz.

Multi-frequency measurements at 950 MHz and 1.8 GHz show that a hilly microcell environment has a larger standard deviation of large scale path loss than a flat environment [41]. For the measurements in [41], the base station antennas range from 9 meters to 30 meters in height and the receiver antenna is fixed at 1.4 meters. The path loss of the two frequencies is closely related. It is reported that there are more areas exceeding a minimum threshold at 1.8 GHz due to the increased diffraction loss. More coverage holes exist at the lower antenna height as well.

Results, for measurements using base station antenna heights of 3.7 meters, 8.5 meters, 13.3 meters for the San Francisco Bay Area are reported in [16]. Path loss exponents ranging between 1.9 and 2.6, with standard deviations between 9.6 and 18.3 dB, are reported. RMS delay spreads range from a worst case value of 549 ns for downtown San Francisco to 1.86 μs in Oakland, CA. It is shown that, in built up areas, the height of the base station antenna does not severely impact the propagation statistics. However, when the surrounding buildings are on the order of the base station antenna height, the propagation statistics depend on the exact antenna height.

Wideband path loss measurements made in San Francisco at 1900 MHz indicated that the first Fresnel Zone is an accurate minimum mean squared error breakpoint for a dual regression path loss model [10], where two different path loss exponents are used before and after the breakpoint. The first Fresnel zone, d_f, is given by

$$d_f = \frac{1}{\lambda}\sqrt{\left(\Sigma^2 - \Delta^2\right) - \left(\Sigma^2 + \Delta^2\right)\left(\frac{\lambda}{2}\right)^2 + \left(\frac{\lambda}{2}\right)^4} \qquad (2)$$

where d_f is the first Fresnel zone distance, $\Sigma = h_t + h_r$, $\Delta = h_t - h_r$, h_t is the transmitter antenna height, h_r is the receiver antenna height and λ is the wavelength.

Table 5 indicates the path loss exponents, standard deviation and break points for a dual regression path loss model having breakpoints at either the first Fresnel zone or at d_b, the optimal break point which minimizes the standard deviation σ [10]. Notice from Table 5 that the Fresnel model using equation (2) provides a standard deviation which is only fractionally larger than the optimum MMSE fit. Also notice the large variation of the path loss exponent n_2 for the MMSE model, and the smaller variation in n_2 when d_f is used as the breakpoint. This clearly illustrates the danger of using arbitrary breakpoints to describe microcell propagation, as a path loss exponent of 9 implies the channel attenuates 90 dB for every decade of distance. Figure 8 shows the measured path loss data as a function of transmitter-receiver separation, and how it fits the dual regression model[10]. Also in [10], measured rms delay spreads were found to be a strong function of the antenna heights, and an exponential upper bound for the worst case rms delay spread was based on path loss.

Table 5: Path loss exponents, standard deviations, and break points for
1st Fresnel zone and MMSE break point curve fits for each antenna height.

	Fresnel Best Fit				MMSE Best-Fit			
	n_1	n_2	σ (dB)	d_f (m)	n_1	n_2	σ (dB)	d_b (m)
Low	2.18	3.29	8.76	159	2.20	9.36	8.64	884
Med	2.17	3.36	7.88	366	2.14	6.87	7.23	884
High	2.07	4.16	8.77	573	2.03	2.79	8.34	190

5 Building penetration of RF waves

Knowledge of path loss between a transmitter outside a building and receivers inside a building will become increasingly important for cordless, paging, PCS, and parasitic cellular systems. Particularly for overlay systems that co-exist in the same frequency band, building penetration will be relied upon to provide RF isolation whenever possible.

Office buildings usually have a steel/reinforced-concrete frame, precast concrete floors and plasterboard internal partition walls [1]. The major obstacles between an outdoor transmitter and an indoor receiver are the walls, floors, furniture and metallic pipes and ventilation ducts. Figure 9 shows the transmission of an electromagnetic wave through a dielectric slab of thickness d. Factors that determine the total attenuation are the reflection at the two interfaces, the attenuation due to the lossy slab medium, and the loss offset due to multiple reflection inside the slab [23]. R is the ratio of reflected to incident power, and is called reflectivity. R is a function of frequency, the incident angle θ, and the electrical properties of the slab. For frequencies above 1GHz, relative permittivity ranging between 2 and 8, conductivity between 10^{-3} and 10^{-2} Siemens, then it follows that the dissipation factor of the slab $\sigma/(\omega\varepsilon)$ << 1, and reflectivity is mainly a function of the incident angle and ε_r of the material. It is observed from measured values [23],[24] that the total transmission loss is approximately constant for θ between 0 and +/- 60 degrees and increase sharply above 60 degrees.

Building penetration loss can be defined in different ways. While Figure 9 demonstrates a basic electromagnetic problem, researchers have defined building penetration loss as the difference between the signal strength measured inside the building and the average signal level measured around the perimeter of the building on the street level [25],[26].

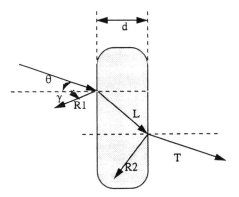

Figure 9. Transmission through a dielectric slab

The problem of modeling radiowave penetration into buildings is three dimensional and in an urban environment may result in a line of sight path for higher floors. The received signal in local areas within buildings is approximately Rayleigh distributed, and the local average signal is approximately lognormally distributed in dB [26]. Work in [35] measured signal strength on different floors of a building with a LOS path to the outdoor transmitter. The results showed that a Suzuki distribution approximated the fading distribution.

Building penetration loss has been shown to depend on numerous factors such as story level, frequency of operation, type of exterior wall, and the surrounding building environment. A few of these factors are discussed in detail below.

5a) Type of exterior wall

Different materials have different transmission and reflection coefficients. Thus, a radio wave propagating through a metallic wall will be attenuated more than the one passing through a brick wall [26]. Consequently, the penetration loss due to metallic walls is greater than that for brick walls. Walls are commonly made of concrete, glass, steelframe, wood or bricks. The building penetration loss for different construction materials can be obtained from [26] and [27]. In [26], the attenuation was found to be greater in metal buildings (23dB) than in buildings with partial metal sides (3dB) or wooden structures (1.5dB) [26]. In [27] building penetration effects in Japan were studied and found to be 26dB for metallic walls. The 3 dB difference in building penetration for metal walls is attributed to the fact that more metal in used in construction of Japanese buildings to protect against earthquakes.

In [28], electromagnetic properties of three common building materials: brick; concrete; and limestone were measured. The measured values of permeability and permittivity of these materials indicate that reflectivity, and consequently the penetration loss, is greatest for concrete, followed by brick and limestone, respectively. Another observation is that there is a variation of these properties with frequency which indicates that the building penetration loss is frequency dependent.

5b) Floor level

The received signal strength from an outdoor transmitter increases with height within a building. This is mainly because the surrounding environment is usually cluttered with buildings (particularly in urban environments) and shadowing effects are more pronounced at the street level than at the top of a high rise building where a line of sight propagation path may be present. This results in a decrease in the building penetration loss with height. The decrease is due to the fact that the penetration loss is defined with respect to the average received signal strength at the street level. In [29] measurements of radio signals into 14 different buildings from 7 different cellular transmitters are described. The results showed that, as compared to the average received signal level at the street, the building penetration loss decreased at the rate of 1.9 dB/floor up to the 15th floor, and started increasing again after the 15th floor. The increase was attributed to the shadowing effects of adjacent buildings. In [30], building penetration loss was measured at the University of Liverpool. The measurements showed that penetration loss decreased at a rate of 2dB per floor up to the ninth floor, above which the penetration loss increased. This was also attributed to shadowing effects of adjacent buildings. Similar behavior was also observed in [31]. In all the above cases the outdoor transmitter height was below the receiver height for the measurements on higher floors where the penetration loss increased. We speculate that the increase in penetration loss at the higher floors observed by all researchers could be attributed to the elevation pattern of the transmitting antennas, which would tend to attenuate the radiation at higher floors.

5c) Frequency dependence

Due to their dielectric nature, the building walls are frequency selective. The building penetration loss shows frequency dependence because of this reason. Measurements made in Liverpool [30] show a decrease in penetration loss with frequency. On average the penetration loss decreased by 1.5 dB from 441 MHz to 896.5 MHz and decreased by 2 to 2.5 dB from 896.5 MHz to 1400 MHz. Measurements indicate a building penetration loss of 16.37, 11.61 and 7.56 dB on the ground floor at frequencies 441, 896.5 and 1400 MHz respectively [30]. Similar observations were also made for frequencies of 900 MHz, 1800 MHz and 2300 MHz [32] where the penetration loss was 14.2 dB, 13.4 dB and 12.8 dB respectively. The variation in building penetration loss with frequency was reported in [24] where the authors found that the penetration loss on the second floor decreased from 35 dB at 200 MHz to 12dB at 600MHz and then increased again to 30 dB at 1 GHz. The frequency selective nature of penetration loss was similar on higher floors, but with a decrease in the actual values of observed penetration loss.

5d) Percentage of windows

When a building has a large number of windows, a lower penetration loss is observed [29] since the waves pass through the windows with little attenuation. Windows may be shielded or unshielded. Shielded windows use a metallic film to reflect sunlight away from the building interior and can exhibit a loss of up to 12 dB. For shielded windows the loss depends upon the shielding material. The loss for copper sputtered windows, for example, was found to be very large and blocked all measurable signals [29]. From [29], windowed areas typically exhibit penetration loss which is 6dB lower than non-windowed areas. Thus, stronger signals will be observed in buildings with more windows

5e) Building surroundings

A multi-story building in an urban area with similar surroundings will experience lower signal strengths inside the building due to the absence of line of sight propagation. In [29], sa building in an urban area exhibited a penetration loss of 18.0 dB, whereas a similar building in a suburban area had a penetration loss of 13.1 dB.

5f) Condition of transmission

The transmission path into a building can be line of sight, partial line of sight, or shadowed where no line of sight propagation is possible. Penetration loss is shown to be different for the three cases [30]. The loss is comparable on all floors for only the line of sight case. For the other two cases, no significant conclusion of path loss variation with floor level could be drawn from the measurements. For line of sight conditions signal strength variations were shown to fit the log-normal distribution with a standard deviation of 4 dB. For partial to complete line of sight conditions the standard deviation was 6-9dB.

6 Conclusion

In this paper, we have presented some of the fundamental challenges facing future wireless communication systems. Interference will exist in the RF channel, and proper models are needed to determine the nature and magnitude of co-channel and adjacent channel interference. Propagation models are not only needed for installation guidelines, but they are a key part of any analysis or design that strives to mitigate RF interference. Also, capacity and system performance predictions are only as good as the channel models they are based upon.

Fortunately, many researchers over the last several years have carried out measurements and have developed propagation models which provide good estimates for signal strength and time dispersion in many new environments which will be served by PCS. We have sought to compile in this paper the state of research of propagation models for emerging wireless systems.

7 References

[1] Rappaport, T.S., "The Wireless Revolution," IEEE Comm. Mag., Nov. 91, pp. 52-71.

[2] Boucher, N.J., The Cellular Radio Handbook, Second Edition, 1992, Quantum Publishing.

[3] Lee, W.C.Y., Mobile Communications Design Fundamentals, Second Edition, 1993.

[4] Chang, J.J.C., et. al, "Wireless Systems and Technologies: An Overview," AT&T Journal, July/Aug., 1993, pp.11-18.

[5] Saleh, A.M., Valenzuela, R.A., "A Statistical model for Indoor Multipath Propagation," IEEE Journal on Selected Areas in Communications, Vol. SAC-5, No.2, Feb. 1987, pp. 128-137.

[6] Hashemi, H., "Impulse Response Modelling of Indoor Radio Propagation Channels," IEEE Journal on Selected Areas in Communications, Vol. 11., No. 7, Sep. 1993, pp. 967-977.

[7] Siedel, S.Y., "Site Specific Prediction for Wireless In-Building Personal Communication System Design," MPRG -TR-93-4, Feb.7,1993, Virginia Tech.

[8] Schaubach, K., "Microcellular Radio Channel Prediction Using Ray Tracing,", MPRG-TR-92-15, Virginia Tech. Also see 1992 IEEE Veh. Tech. Conf. Proceedings.

[9] Bertoni, H.L., Honcharenko, W., Leandro, R.M., Xia, H.H., "UHF Propagation Prediction for Wireless Personal Communications" to appear in IEEE Proceedings.

[10] Blackard,K.L., et. al., "Path Loss and Delay Spread Models as Functions of Antenna Height for Microcellular System Design," 1992 IEEE Veh. Tech. Conf., Denver, CO, May 1992.

[11] Yacoub, M.D., Foundations of Mobile Radio Engineering, CRC Press Boca Raton, 1993.

[12] Seidel, S.Y. and Rappaport, T.S., "914 MHz Path Loss Prediction Models for Indoor Wireless Communications in Multifloored Buildings," IEEE Trans. on Ant. and Prop., Vol. 40, No. 2, Feb. 1992, pp. 1-11.

[13] Rappaport, T.S., "Characterization of UHF Multipath Radio Channels in Factory Buildings," IEEE Trans. on Antennas and Prop., Vol. 37, No. 8, Aug. 1989, pp. 1058-1069.

[14] Pahlavan, K., Ganesh, R. and Hotaling, T., "Multipath Propagation Measurements on Manufacturing Floors at 910 MHz," Electronics Letters, 2nd Feb., 1989, Vol. 25, No. 3, pp. 225-227.

[15] Hawbaker, D. A., "Indoor Wide Band Radio Wave Propagation Measurements and Models at 1.3 GHz and 4.0 GHz," Electronics Letters, Vol. 26, No. 21, 11th Oct., 1990, pp. 1800-1802.

[16] Telesis Technologies Laboratory, Experimental Licence Report to the FCC, Aug. 1991.

[17] Devasirvatham, D.M.J., "A Comparison of Time Delay Spread and Signal Level Measurements of 850 MHz Radio Waves in Building Environments," IEEE Trans. on Ant. and Prop., Vol. AP-35, No. 3, March 1987, pp. 319-324.

[18] Bultitude,R.J.C., "Measurement Characterization and Modelling of Indoor 800/900 MHz Radio Channels for Digital Communications," IEEE Comm. Mag., June 1987, Vol. 25, No. 6.

[19] Parsons, J.D., The Mobile Radio Propagation Channel, Halsted Press, N.Y., 1992.

[20] Walfisch, J., Bertoni, H. L., "A theoretical model of UHF propagation in urban environments," IEEE Trans. Antennas and Propagation, Vol. 36, No. 12, 1988, pp. 1788-96.

[21] Hashemi, H., "Impulse Response Modeling of Indoor Radio Propagation Channels," IEEE J. on Sel. Areas of Comm., Vol.11, No.7, Sep., 1993, pp-967-978.

[22] Bultitude, R.J.C., et. al., "The Dependence of Indoor Radio Channel Multipath Characteristics on Transmit/Receive Ranges," IEEE J. on Sel. Areas in Comm, Vol. 11, No.7, Sep., 1993, pp. 979-990.

[23] Tang, Y., Sobol, H., "Microwave Propagation in Multi-room Buildings for PCS," IEEE GLOBECOM 1993, Vol. 2, pp. 1232-1236.

[24] Macario, R. C. V., "How building penetration loss varies with frequency," IEEE Vehicular Technology Society Newsletter, Nov. 1993, pp. 26-27.

[25] Rice, L.P., "Radio Transmission Into Buildings at 35 and 150 MHz," Bell Sys. Tech. J., 1959, 38(1), pp. 197-210.

[26] Cox, D.C., Murray, R.R and Norris, A.W., "Measurement of 800 MHz Radio Transmission Into Buildings With Metallic Walls," Bell Sys. Tech. J., 1983 (62) pp. 2695-2717.

[27] Kozono, S., and Watanabe, K., "Influence of Environment Buildings on UHF Land Mobile Radio Propagation," IEEE Trans. Comm., Com-25, Oct. 1977, pp. 1133-1143.

[28] Sou, C. K., Landron, O., Feuerstein, M. J., "Characterization of Electromagnetic Properties of Building Materials for use in Site-Specific Propagation Prediction," MPRG-TR-92-12, June 1992. Also see paper by Landron in 1993 IEEE Veh. Tech. Conf. Proceedings.

[29] Walker, E.H., "Penetration of Radio Signals Into Buildings in Cellular Radio Environment," Bell Sys. Tech. J., 1983, 62, (9), pp. 2719-2735.

[30] Turkmani, A.M.D., Parsons, J.D, Lewis, D.G., "Radio Propagation Into Buildings at 441, 900 and 1400 MHz.," Proc. of Fourth International Conference on Land Mobile Radio, Dec. 1987, pp.129-138.

[31] Durante, J.M., "Building Penetration loss at 900 MHz", Conf. Proc., IEEE Veh. Tech. Conf., 1973, pp. 1-7.

[32] Turkmani, A.M.D., and Toledo, A.F., "Propagation Into and Within Buildings at 900, 1800 and 2300 MHz.," 1992, 42nd IEEE Vehicular Tech. Conf., pp. 663.

[33] Horikishi, J., Tanaka, K., and Morinaga, T., "1.2 GHz. Band Wave Propagation Measurements in Concrete Buildings for Indoor Radio Communications," IEEE Trans. of Veh. Tech., 1986, VT-35 (4), pp. 146-152.

[34] Molkdar, D., "Review on radio propagation into and within buildings," IEE Proc., Vol. 138, Feb. 1991, No. 1.

[35] Barry, P.J. and Williamson., A.G., "Modelling of UHF radiowave signals within externally illuminated multi-storey buildings," J.IERE, Vol. 57, No. 6, Dec., 1987, pp. S230-S240.

[36] Chia, S.T.S., "1700 Mhz Urban Microcells and Their Coverage into Buildings," IEE Conference Publication 1980, pp. 504-507.

[37] Honcharenko,W., Bertoni, H.L., "Mechanisms Governing Propagation Between Different Floors in Buildings," IEEE Trans. on Ant. & Prop., Vol.41, No.6, June 1993, pp.787-790.

[38] Blackard, K.L., Rappaport, T.S., Bostian, C., "Measurements and Models of Radio Frequency Impusive Noise for Indoor Wireless Communications," IEEE J. on Sel. Areas in Comm., Vol. 11, No. 7, Sep. 1993, pp. 991-1001.

[39] Harley, P., "Short Distance Attenuation Measurements at 900 MHz and 1.8 GHz Using Low Antenna Heights for Microcells," IEEE J.Sel. Areas Comm., Vol.7, No.1, Jan., 1989, pp. 5-11.

[40] Green, E., "Path Loss and Signal Variability Analysis for microcells," IEE 5th International Conference on Mobile Radio and Personal Communications, Warwick, UK, Dec. 11-14, 1989, pp. 38-42.

[41] Pielou, J.M., "Preliminary Analysis of 1.8 GHz Measurements for Personal Communications," IEE Conf. on Ant. and Prop., U.K, April 1991, pp. 508-511.

[42] Panjwani, M.A., et. al., "An Interactive System for Visualizing Wireless Communication Coverage Within Buildings," 4th Virginia Tech. Symposium on Wireless Personal Comm., Blacksburg, Virginia, June 1-3, 1994.

[43] Rappaport, T.S., Seidel,S., Takamizawa,K., "Statistical Channel Impulse Response Models for Factory and Open Plan Buildings Radio Communication system Design" IEEE Trans. on Comm, May 1991.

[44] Bultitude, J.C., Mahmoud, S.A., and Sullivan, W.A., "A Comparison of Indoor Radio Propagation Characteristics at 910 MHz and 1.75 GHz," *IEEE Journal on Selected Areas in Communications*, Vol. 7, No. 1, January 1989.

[45] McCulley, S.L., Rappaport, T.S., "DSP Techniques for Cellular, Paging, and PCS Intercept," 1993 Tactical Technologies and Wide Area Survillance International Symposium, Chicago, IL Nov. 2-5, 1993, pp. 185-189.

[46] Seidel, S.Y, Rappaport, T.S., "A Ray Tracing Technique to Predict Path Loss and Delay Spread Inside Buildings," IEEE Globecom ' 92, Dec. 7, 1992, pp.1825-1829.

[47] Ho, C.M.P., Rappaport, T.S., Koushik, P., "Antenna Effects on Indoor Obstructed Wireless Channels and a Deterministic Image-Based Wide-Band Propagation Model for In-Building Personal Communication Systems," Intnl. J. of Wireless Info. Networks, Vol.1, No.1 1994, pp.61-76.

[48] Valenzuela, R.A., "A Ray Tracing Approach to Predicting Indoor Wireless Transmission," IEEE Veh. Tech. Conf., 1993, pp.214-218.

2

A Survey of Adaptive Single Channel Interference Rejection Techniques for Wireless Communications

J. D. Laster and J. H. Reed

Mobile and Portable Radio Research Group
Bradley Department of Electrical Engineering,
Virginia Polytechnic Institute & State University
Blacksburg, Virginia 24061-0111

ABSTRACT

Tremendous growth in wireless communications has greatly increased loading of the spectrum. Spectrum loading translates into a higher likelihood of users interfering with one another. Interference rejection techniques provide a means of minimizing this multi-user interference, allowing greater usage of available spectrum.

This survey paper focuses on single-channel techniques for interference rejection (that is, techniques employing one antenna) as opposed to multi-channel techniques (which utilize arrays or cross-polarized antennas). Implementation papers are deemphasized in this survey since techniques constitute the main interest. The paper divides interference rejection techniques for digital modulation into spread spectrum techniques and non-spread spectrum techniques.

Spread spectrum categories include direct sequence (DS), code division multiple access (CDMA), and frequency hopping (FH). DS techniques focus on rejection of narrowband interference and include whitening filters (i.e., adaptive notch filters), decision feedback filters, and adaptive A/D conversion. Some CDMA techniques employ interference estimation-and-subtraction from the signal-of-interest, while others exploit spectral correlation properties. FH techniques apply whitening filters and make use of the transient nature of the hopping signal. Non-spread spectrum techniques include the constant modulus algorithm, neural networks, non-linear filters, and time-varying filters that use spectral correlation properties.

1   INTRODUCTION

The growth in wireless communications necessitates more efficient utilization of spectrum. The increased sharing of spectrum translates into a higher likelihood of users interfering with one another. Interference rejection techniques provide a means of minimizing this multi-user interference, allowing greater usage of available spectrum.

Interference rejection is important for several reasons. As newer communication technologies supersede older technologies, interference rejection techniques are important in helping to facilitate compatibility during transitions between the old and new technologies. Several examples illustrate the need for compatibility: co-utilization of the existing cellular band with new narrowband code division multiple access (CDMA) and time division multiple access (TDMA) digital cellular signals, broadband CDMA overlaying AMPS signals in the cellular bands, co-utilization of the new personal communication system band (1.8 - 2.2 GHz) with existing microwave systems, the addition of a vast number of new low-earth-orbiting (LEO) satellites with overlapping footprints with older satellites, accomodation of high definition television (HDTV) transmisssions within the current TV band, and new commercial digital transmissions in the FM broadcast band.

The CDMA/AMPS interference problem is the key problem in making viable a new digital cellular format. Schilling, Lomp, and Garodnick [1] (1993) present a broadband CDMA (B-CDMA) scheme that will overlay the existing cellular telephone spectrum (825-894 MHz). The overlay will provide additional capacity to the network while allowing high quality voice and high speed data services to be coexistent with the existing cellular services (AMPS and TDMA). The absence of mutual interference to and from the B-CDMA overlay is accomplished using an adaptive filter.

In satellite-based personal communications systems, geostationary satellites can interfere with each other, as well as with LEO satellites, limiting capacity. This issue is especially relevant because of the large number of LEO satellites proposed for worldwide cellular and information networks. An informative overview of satellite interference is found in the work of Kennedy and Koh [2] (1981). Their paper discusses the background and relevance of the problem of frequency-reuse interference in TDMA/QPSK satellite systems and discuss techniques to alleviate interference effects.

The military applications of interference rejection are numerous. The most obvious application is in mitigating the effects of intentional jamming. A not so obvious application is the mitigation of self-jamming from harmonics produced by operating transmitters and receivers in close proximity to each other

[3]. In addition, in reconnaissance applications, a stand-off receiver covering a wide geographical region is subject to interference from non-intelligence bearing signals operating in the same band.

This survey paper focuses on single-channel techniques for interference rejection (that is, techniques employing one antenna) as opposed to multi-channel techniques (which employ multiple antennas, such as arrays or cross-polarized antennas). The military has always been interested in single channel techniques because they are generally cheaper, less complex, smaller in size, and more suited to rugged military applications than multi-channel techniques. Along the same lines, the commercial wireless community will likely favor interference rejection techniques which are inexpensive and simple to implement. Implementation papers are deemphasized in this survey since techniques constitute the main interest. The survey focuses on papers published dating from 1980 to the present. The paper divides interference rejection techniques for digital modulation into spread spectrum techniques and non-spread spectrum techniques.

2   SPREAD SPECTRUM TECHNIQUES

Spread spectrum (SS), by its very nature, is an interference tolerant modulation. However, there are situations where the processing gain is inadequate and interference rejection techniques must be employed. This is especially true for direct sequence spread spectrum (DS SS) which suffers from the near-far problem. For this survey, spread spectrum categories include direct sequence (DS), code division multiple access (CDMA), and frequency hopping (FH). Two tutorial papers of particular interest include one by Kohno [4] (1991) and one by Milstein [5] (1988). Kohno provides an overview of classic solutions and promising techniques being studied in Japan. In particular, he describes a temporal domain approach where an adaptive digital filter (ADF) is employed to adaptively identify the time-varying response of the co-channel interference in DS SS multiple access (MA) without excessive noise enhancement. Milstein discusses in depth two classes of rejection schemes (both of which implement an adaptive notch filter or whitening technique): 1) those based upon least mean square (LMS) estimation techniques, and 2) those based upon transform domain processing structures. The improvement achieved by these techniques is subject to the constraint that the interference be relatively narrowband with respect to the DS waveform.

32

2.1 Interference Rejection for Direct Sequence

Interference rejection techniques for DS SS systems are numerous. In particularly, much literature exists on the whitening filter, or adaptive notch filter, as it relates to rejecting narrowband interference (NBI) in DS SS. Decision-directed adaptive filtering is another well established technique for interference rejection. Emerging techniques for narrowband DS SS are adaptive analog-to-digital (A/D) conversion and non-linear filtering.

2.1.1 Whitening Filter or Adaptive Notch Filter

The basic idea in employing a whitening filter is to flatten the spectrum of the signal and interference. SS tends to have a flat and wide spectrum and is affected little by the whitening process, while NBI is characterized by spikes in that spectrum. The whitening filter places notches at the location of the NBI to bring the interference level down to the level of the SS signal. Usually the whitening filter is implemented by an adaptive filter configured as a predictor of the narrowband signal. The wideband SS signal appears at the error output of the adaptive filter.

Saulnier [6] (1992) investigates the use of transform-domain adaptive filtering for the suppression of narrowband jammers in a DS SS receiver. An input signal consisting of the DS signal, white Gaussian noise, and a narrowband jammer, or narrowband SS signal, is transformed into the frequency domain where the adaptive filter acts to suppress the jammer. The key idea behind this technique is to use weight leakage to suppress the interference while preserving the desired DS signal.

Compernolle and Van Gerven [7] (1992) extend the classical LMS noise canceler for the case of signal leakage into the noise reference. The algorithm is derived intuitively from the interpretation of the adaptive noise canceler as a decorrelator between signal estimate and noise, in which the noise reference is replaced by a signal-free noise estimate. Thus, a symmetric adaptive decorrelator is obtained for signal separation. Theoretically, there are many problems concerning stability and convergence, but the authors show that these restrictions are rarely violated in practical applications.

Doherty [8;9;10] (1993;1992) presents a constrained LMS technique that utilizes a constrained optimality criterion to enhance the detection capabilities of DS SS systems. Two transversal tapped delay lines (TDLs) are operated simultaneously, one containing the received data and the other containing the constraint data, as one set of adaptive weights operates on both TDLs with the LMS algorithm as the update technique. The filter weights are updated with respect to both, minimizing the mean-square output error and minimizing the constraint error, with two types of constraint conditions: a correlation-matching condition and a minimum-filter-energy condition. Doherty also proposes an adaptive technique that utilizes the known characteristics of the pseudo-noise (PN) SS sequence to enhance the detection capabilities diminished by interference excision. In [11] Doherty (1991) presents an enhancement of the whitening filter technique which adds constraints based on the known characteristics of the PN SS sequence to enhance the detection capabilities diminished by interference excision. Operating without training bits, the constrained updating of the filter coefficients retains the interference rejection properties of the excision filter while decreasing the variance of the decision variable.

Gevargiz, Das, and Milstein [12] (1989) demonstrate the advantage of an intercept receiver which uses a transform-domain-processing filter and detects DS BPSK SS signals in the presence of NBI by employing adaptive NBI rejection techniques. The receiver uses one of two transform-domain-processing techniques. In the first technique, the NBI is detected and excised in the transform domain by using an adaptive notch filter. In the second technique, the interference is suppressed using soft-limiting in the transform domain. The authors [13;14] (1986) also give an implementation of a transform domain processing radiometer for DS SS signals with an adaptive narrowband interference exciser.

Bershad [15] (1988) investigates the effects of the LMS adaptive line enhancer (ALE) weight misadjustment errors on the bit error rate (BER) for a DS SS binary communication system in the presence of strong NBI. The converged ALE weights are modeled as the parallel connection of a deterministic FIR (finite impulse response) filter and a random FIR filter. The statistics of the random filter are derived, assuming the output of the random filter to be primarily due to the jammer convolved with random filter weights, yielding a non-Gaussian output which causes significant error rate degradation in comparison to a Gaussian model.

Stojanovic, Dukic, and Stojanovic [16] (1987) use linear mean-square estimation to determine the tap weights of two-sided adaptive transversal filters so as to minimize the receiver output mean-square error caused by the presence of NBI and the additive white Gaussian noise (AWGN). The results obtained show a significant reduction of the error rate in comparison to previously published results.

He, Lei, Das, and Saulnier [17] (1988) discuss the modified LMS algorithm for transversal filter structures and lattice filter structures, comparing their BER performance and convergence characteristics.

Lee and Lee [18;19] (1987) suggest a lattice gradient-search fast converging algorithm (GFC). For the case of a sudden parameter jump or new interference, the transient behavior of the receiver using a GFC adaptive lattice filter is investigated and compared with those of the receiver using a LMS or a lattice adaptive filter. The use of a lattice filter is also explored by Saulnier [20;21], Yum, and Das (1987). Here, the reflection coefficients of the filter are adapted using an iterative, steepest-descent adaptive lattice algorithm and a gradient descent algorithm in an effort to minimize the mean-squared error, and in the process, suppress NBI. Adaptive lattice filtering has several advantages: low sensitivity to round-off error and parameter perturbation, decoupling of filter stages, Gram-Schmidt orthogonalization of reverse error residuals, and convergence time independent of the data's spectral dynamic range. Guilford and Das [22] (1985) present an adaptive lattice filter structure as a solution to the narrowband jammer rejection problem in a DS SS communication system, discussing both the adaptive and nonadaptive lattice filters. Results show that even a simple (two-stage) adaptive lattice filter significantly improves probability of error in a DS SS receiver that has been corrupted by NBI.

Saulnier, Das, and Milstein [23] (1985) extend their earlier work [24] (1984) on the digital implementation of the LMS algorithm which uses a burst processing technique to obtain some hardware simplification. While, in general, the results show significant improvement in performance when the LMS algorithm is used, performance declines quickly with increased jammer bandwidth. The authors [25] (1985) also discuss an LMS-based jammer suppression structure which does not require carrier synchronization.

Bandy and Krause [26] (1985) investigate three different methods for implementing a whitening filter called the sequential least-squares autoregressive method, the linear prediction method, and the periodogram method. In addition, Baier and Friederichs [27] (1985) present a technique for canceling wideband interference.

Fox, MacMullen, and Nauda [28] (1985) discuss the design of a frequency domain adaptive filter for the maximum likelihood demodulation of keyed angle-modulated signals in the presence of colored Gaussian interference. When significant errors exist in the estimation of the signal parameters (such as in amplitude estimation), a Wiener filter has been determined to provide a better BER performance for a class of angle-modulated signals over that of a whitening filter.

Mammela [29] (1988) simulates the performance of optimal and adaptive interference suppression filters for DS SS systems. The simulations include the linear M-step prediction and interpolation filters and some of the best-known iterative and time-recursive algorithms (LMS, Burg, and Kalman algorithms). Linear filters are shown to work well if the interference bandwidth is a small fraction of the signal bandwidth. The linear interpolation filter works better than the prediction filter.

Lindquist, Severance, and Rozran [30] (1984) present the ALE and the adaptive noise canceler (ANC) using various algorithms. They develop these algorithms in the frequency domain for efficient implementation.

Plotkin and Wulich [31] (1982) compare two LMS adaptive algorithms: (1) using the conventional LMS notch filter (LMS-NF) and (2) using the function elimination notch filter (FEF-NF). LMS-NF is based on the conventional linear prediction procedure. Superior to LMS-NF, FEF-NF uses *a priori* knowledge of the structural properties of the canceled signal and tunes only part of the filter coefficients. Plotkin and Plotkin [32] (1982) consider the problem of impulsive noise suppression by using an adaptive prediction technique and propose the Time Domain Notch Filter (TD-NF) structure and a procedure to estimate the filter coefficients. The obtained results confirm that the proposed linear algorithm enables a transmission in a background of high power impulsive interference.

Ketchum and Proakis [33] (1982) present results of an investigation of algorithms for estimating and suppressing NBI in PN SS. Techniques for determining the coefficients of a linear, interference suppression filter are described, which are based on linear prediction and conventional spectral analysis methods.

Li and Milstein [34] (Dec. 1981) examine prediction error filters and adaptive transversal filters with two-sided taps. If the frequency of the jamming tone is near the carrier frequency of a PSK signal, the performance of the transversal filter with two-sided taps is better than that of the prediction error filter with the same number of taps. Furthermore, the transversal filter with two-sided taps has the advantage of simplicity due to symmetry.

Dunlop and Whitworth [35] (April 1980) adjust the transmitted signal spectrum by operating on the pseudo-random bit stream with a transversal digital filter before transmission. A similar adaptive filter is used in the receiver pseudo-random code generator, and the demodulated output is obtained by correlation between the two modified code signals. Rejection of NBI is achieved by notching the reference code spectrum at the appropriate frequencies.

2.1.2   Decision Feedback

Decision feedback (DF) techniques use an adaptive filter to notch interference. Decisions (or "best guesses" of the signal state) are made at the output of the filter and then fed back to train the adaptive filter and/or be included in the filtering process. Variations of this technique exist where either the incoming signal is filtered and/or the estimation error is filtered.

Pateros and Saulnier [36;37] (1992) analyze the BER performance of the adaptive correlator in a complex channel with both multipath and NBI. The adaptive correlator, using DF, is shown to be capable of removing relatively wideband interference in the transmission bandwidth.

Ogawa, Sasase, and Mori [38] (1991) examine the performance of a differential phase-shift-keying (DPSK) DS SS receiver using DF filters in the presence of NBI and multipath. They found that the two-sided DF filter is superior for suppressing both interference and multipath in the SS system. Miyagi, Ogawa, Sasase, and Mori [39] (1990) analyze the performance of three types of quadrature partial response signaling (QPRS) systems using complex one-sided and two-sided transversal filters, with additional DF taps, in the presence of single continuous wave (CW) interference and AWGN. They find that both DF filters can suppress not only CW interference, but also filtered noise, while decoding binary data signals from ternary QPRS signals. They also show that the duobinary system has the best performance of the three types of QPRS systems when the frequency of CW interference is low. Similarly, Ogawa, Sasase, and Mori [40] (1989) examine suppression of CW interference and colored noise in a QPSK system using DF filters.

Dukic, Stojanovic, and Stojanovic [41;42;43] (1990;1989;1987) include two-sided transversal filters along with DF to combat NBI. Their receiver is made up of two branches: the conventional demodulator followed by a DF filter and in an auxiliary branch, a demodulator with the carrier in quadrature followed by a two-sided adaptive transversal filter. The results show that NBI rejection is fairly high and practically does not depend upon the difference of frequencies of the desired and interfering carriers or upon the interfering carrier level. In [44], Dukic, Cuberovic, Stojanovic, and Stojanovic (1990) give an extended analysis of their work.

Shah and Saulnier [45] (1988) conclude that LMS adaptive filtering improves the probability-of-error performance of a DS SS system operating in the presence of stationary single-tone jammers. They also claim that when compared with the no feedback case, LMS adaptive systems with DF do not degrade probability of error performance; however, DF does not always appreciably improve system error rates either.

Liu and Li [46] (1987) examine the performance of QPSK systems using adaptive DF filters in the presence of finite-bandwidth interference and show that the performance of filters with DF is better than that of the filters without DF. They model finite-bandwidth interference as a first order autoregressive process. Under the assumption of perfect DF, Takawira and Milstein [47] (1986) develop analytical expressions for the bit-error rates. They compare two different DF filter realizations to show the advantage of filtering the interference and noise only, excluding the desired signal. Li and Milstein [48] (1983) analyze the performance of QPSK systems using one-sided (with lagging taps) and two-sided transversal filters with additional DF taps, in the presence of Gaussian noise and a single CW interferer. They show that if the filter is used for rejecting CW interference only, the one-sided DF filter is preferred.

Libing, Guangguo, and Boxiu [49] (1991) analyze the performance of QAM systems using adaptive transversal filters with additional DF taps, in the presence of Gaussian noise and FDM-FM interfering signals. The LMS algorithm is used to update the tap weights.

2.1.3 Adaptive A/D Conversion

The idea behind adaptive A/D conversion is that the bias introduced by a high-power narrowband interferer can be tracked and compensated before entering the A/D converter. Equivalently, thresholds of the A/D converter can be changed to minimize the impact of the interference.

Goiser and Sust [50;51] (1990) consider digital matched filters for DS SS communications and found that minimum complexity is obtained if hard-limiting analog-to-digital converters (ADCs) are used. Performance of a 2-bit ADC (with adaptive thresholds) and hard limiting receiver in AWGN, CW, and combined CW/AWGN interference is evaluated. Amoroso [52] (1984) extends previous analyses to give the performance of the adaptive 2-bit A/D converter in combined CW and Gaussian interference. The converter is seen to give substantial conversion gain even when the DS SS is much weaker than the Gaussian component of interference. The upper bound on conversion gain depends primarily on the relative strengths of the Gaussian and CW components of interference.

Another technique for making the DS receiver more robust with respect to interference [53;54;55] (1983;1986;1987) is a scheme using an A/D converter, in conjunction with a variable threshold, to retain those chips of the spreading sequence which, when added to a strong interfering signal, are still received with their correct polarity.

2.2 Interference Rejection for Code Division Multiple Access

The current generation of CDMA systems employ single stage correlation receivers that correlate the received signal with a synchronized copy of the desired signal's spreading code. In a single cell environment, CDMA systems employing simple correlation receivers cannot approach the spectral efficiency of time or frequency division multiple access[56]. Furthermore, correlation receivers are particularly susceptible to the near-far problem when multiple access signals are received with different signal powers. Even if sophisticated power control is employed, the near-far effect can result in significant performance degradation. Greater channel capacity for CDMA can be achieved by using interference rejection techniques. Much of the motivation for this work has resulted from the theoretical capacity work by Verdu [57] (1986) and Lupas [58] (1986) for optimal CDMA receivers. Because this receiver is hopelessly complex, a back-propagation neural net has been used by Aazhang, Paris, and Orsak [59] (1992) to approximate the receiver.

There are two categories of CDMA interference rejection techniques. The first of these categories encompasses cyclostationarity exploiting algorithms which, in many instances, resemble fractionally-spaced equalizers (FSEs). The second class of techniques is based on estimating and subtracting the multi-access interference. The cyclostationarity exploiting algorithms are described first.

Monogioudis, Tafazolli, and Evans [60] (1993) employ a technique based on adaptive linear fractional spaced equalization (LFSE) to adaptively cancel the multiple-access interference in CDMA systems. The simulation results indicate that the LFSE offers significant gains over the conventional detector, eliminating the near-far problem without the explicit knowledge of the interfering spreading sequences.

Davis and Milstein [61] (1993) describe an adaptive tapped delay line equalizer which operates in a DS-CDMA receiver. This structure, without modification, can be applied to make the system insensitive to spectral inefficiency caused by non-ideal pulse shaping and coherently combined multipath to reject NBI. The technique is compared to previously published methods of NBI rejection.

Rapajic and Vucetic [62] (1993) describe a fully asynchronous single user receiver in a CDMA system where the receiver is trained by a known training sequence prior to data transmission, and continuously adjusted by an adaptive algorithm during data transmission. An adaptive, fractionally-spaced LMS filter, instead of matched filters with constant coefficient,s is employed for each user separately. Experimental results show that a considerable improvement in BER is achieved with respect to the conventional single-user receiver. Abjulrahman, Falconer, and Sheikh [63] (1992) present related work on fractionally-spaced DF minimum mean squared error (MMSE) filters in the context of DS SS CDMA. They show its performance in slow fading and how an FSE can be used as a CDMA demodulator even when slow fading is present. The advantage of this receiver is its simplicity.

Madhow and Honig [64;65] (1992) consider interference suppression schemes for DS SS CDMA systems using the MMSE criterion. These schemes have the virtue of being amenable to adaptation and are simple to implement, while, at the same time, alleviating the near-far problem to a large extent. The channel output is first passed through a filter matched to the chip waveform and then sampled at the chip rate. Because of the complexity and coefficient noise associated with such an adaptive filter when N is large, simpler structures with fewer adaptive components are proposed. In each case the multiple samples per symbol are combined via a tapped delay line, where the taps are selected to minimize the mean square error.

Holley and Reed [66] (1993) and also Aue and Reed [67] (1994) show how spectral correlation properties can be exploited by a time-dependent adaptive filter to provide spectral capacity for CDMA close to that of FDMA or TDMA using frequency-domain and time-domain filtering structures.

Asynchronous CDMA systems using digital matched filtering (DMF) reception techniques suffer from poor multiple access spectral efficiency. This is due entirely to the lower bounds on the mean square cross-correlation levels which exist between the orthogonal codes allocated to each system subscriber. This interference can be estimated through channel measurement, and it is then possible to regenerate and subsequently cancel cross-correlation components from individual interfering channels.

Multi-stage correlation receivers, due to Varansi and Aazhang [68] (1990), are a promising suboptimal approach for reducing interference and increasing channel capacity. At each stage, a bank of single-user receivers demodulates the received signal. After each stage, the estimated signals for all interference sources are subtracted from the received signal, and then demodulation is repeated. This procedure can be repeated for an arbitrary number of stages to obtain an iterative estimate of the interference. Simulation results for a simple channel model indicate that only a few stages are necessary to achieve most of the potential performance improvement.[69]

Most work on multistage interference rejection techniques has focused on idealized Gaussian channels and assumes perfect knowledge of each user's power. In order for multi-stage receivers to perform well on the time-varying multipath channels which are characteristic of the wireless environment, they must simultaneously form an estimate of channel characteristics and interference characteristics. Fortunately, the use of a RAKE receiver (or similar adaptive channel estimator) in a multi-stage algorithm makes this possible. The RAKE can be used to simultaneously estimate the received data and the channel impulse response for each user. The estimated signal will then be reconstructed and subtracted from the received signal. The new received signal is used to perform another stage of estimation. This approach allows the multi-stage receiver to form iterative estimates of both the channel and the received data, and allows the multi-stage receiver to operate with no prior information about channel conditions. Using this structure, Kawabe, et al. [70] (1993), in particular, show a spectral efficiency increase by a factor of eight. Several research groups are investigating similar receiver structures [71;72;73;74]. Mowbray, Pringle, and Grant [75;76] (1992) show an upper bound on the spectral efficiency approaching 130% or 1.3 normalized channels per hertz for successive cascaded cancellation stages.

2.3 Interference Rejection for Frequency Hopping

Frequency hopping interference rejection techniques often employ a whitening stage to reject narrowband and wideband interference. In some instances, they also use the transient property of the hopper to distinguish it from persistent background interference.

Glisic and Pajkovic [77;78] (1991;1990) analyze the performance of a DS QPSK SS receiver using adaptive filtering to reject a FH MA signal. Considering the adaptive prediction error filter with two-sided taps, they show graphically the conditions and number of FH MA signals that can be efficiently suppressed using adaptive filtering in a DS SS receiver.

Unlike DS signals, FH signals are instantaneously narrowband, but when observed over a time span encompassing multiple hops, the FH signal becomes wideband. Exploiting this property, Iltis [79] (1986) shows how prewhitening filters designed using linear least-squares estimation techniques can be applied to improve the detection performance of FH signals. One filter, with taps spaced at the hop duration T_h, can reject interference with a bandwidth of up to π/T_h radians/sec. A second filter uses "fractionally-spaced" taps at intervals of T_h/L (where L is the number of hops) and rejects interference with a bandwidth of up to $L\pi/T_h$ radians/sec, providing improved detection performance when the FH signal is linearly combined over L hops.

Reed and Agee [80] (1992) extend the idea of whitening using a time dependent filter structure to estimate and remove the interference based on the interference spectral correlation properties. The process can be viewed as a time-dependent whitening process with suppression of signals that exhibit a particular spectral correlation. The detectability of FH SS in the presence of spectrally correlated interference is nearly independent of the SIR.

Bishop and Leahy [81] (1985) present a technique for enhancing a wideband signal of narrow instantaneous bandwidth, such as a FH signal, from wideband and NBI. The central concept is that statistical estimation inherently involves a time average, with an accompanying convergence time, and this property can be used to separate signals. A device, such as an ALE that separates wideband and narrowband waveforms, can use this property to distinguish the signal-of-interest (SOI) from interference.

Gulliver [82] (1991) proposes a concatenation of order statistics (OS) and normalized envelope detection (NED) to combat noise and multitone jamming. He shows that the OS-NED method significantly improves the performance of NED in multitone jamming, with only slight degradation in noise jamming.

2.4 Novel Interference Rejection for Spread Spectrum

Garth and Poor [83] (1992) develop DS SS suppression algorithms which are based on non-linear filters that produce predictions of the interfering signals that are then subtracted from the received signal to suppress the interference. They show the independence of the filtering procedures from the noise distribution shape, given constant second-order statistics. Garth, Vijayan, and Poor [84] (1991) generalize the non-linear filter derived by Poor and Vijayan [85] (1990) and show that for channels corrupted by impulsive noise, the binary nature of the DS signals can be exploited to obtain performance using non-linear filters. Overall, the interference rejection capability provided by the non-linear filter over the linear for impulsive noise background is substantial.

Dominique and Joseph [86] (1988) propose the use of a non-linear transform domain modified median filter (MMF) that removes only the impulse-like jammers and not the desired signal component. The MMF does not require any convergence time, has a fixed processing delay, and requires no *a priori* jammer information. The technique can be categorized as nonparametric interference suppression.

Higbie [87] (1988) describes a non-linear signal processing technique designed to suppress interference in DS SS receiving systems. The basic idea is to optimize the detection process dynamically, in the presence of current interference, by estimating the statistics of the interference and then by using this information to derive a non-linear transform to apply to the corrupted signal. This adaptation is open-loop, thus avoiding convergence problems and yields large improvements (tens of dB).

3 NON-SPREAD SPECTRUM TECHNIQUES

A number of techniques exist for rejecting interference for non-spread spectrum signals. Many of these techniques, such as the constant modulus algorithm and decision-directed adaptive filtering are well known equalization techniques. In addition, some emerging interference rejection techniques which are based on neural nets, time-dependent filtering, and non-linear filtering show great promise.

3.1 Constant Modulus Algorithm

Interference and channel distortion will alter the envelope of a constant modulus (envelope) signal. The constant modulus algorithm (CMA) works by adapting a filter to restore the constant envelope, thereby rejecting interference and suppressing channel distortion. Most of the literature concerning the CMA focuses on its equalization capability. Here, literature is addressed which concerns the CMA's interference rejection capability.

Kwon, Un, and Lee [88] (1992) investigate the convergence properties of the CMA when applied to interference rejection. They also analyze the convergence behavior of the squared output modulus and the MSE of the modulus. They found that the convergence behavior can be modeled by a recursive equation with a varying convergence factor.

Rude and Griffiths [89] (1990) develop a fractionally-spaced adaptive equalizer based on the linearly constrained constant-modulus (LCCM) algorithm. The LCCM algorithm exploits prior knowledge of synchronization, sampling strategy, and pulse shape to prevent capture of the constant-modulus (CM) algorithm by narrowband constant envelope interferers. LCCM uses *a priori* knowledge of only the SOI. Simulations show that this approach greatly reduces the vulnerability of CMA to strong constant envelope interferers and yields a set of tap values that can be successfully used as initial conditions for follow-on DF adaptation.

Ferrara [90] (1985) also presents a method for adaptively canceling interference from a constant envelope target signal, even when some of the interfering signals are also constant envelope. The adaptive algorithm distinguishes between target signal and interference on the basis of signal amplitude and envelope shape, given that the amplitude of the target signal is approximately known or measurable.

Gooch and Daellenbach [91] (1989) describe a technique for preventing interference capture by using a spectral whitening algorithm to initialize the filter weights prior to switching to the CMA. The method requires no knowledge of the received interference scenario, and it allows notching of one or more interferers. Satorius, et al., [92] (1988) compare the interference rejection performance of the CMA to linear prediction or whitening techniques.

3.2 Neural Networks

Howitt, Reed, Vemuri, and Hsia [93] (1993) survey recent developments in applying neural nets to equalization and interference rejection. Advantages of neural nets over conventional linear filtering and equalization include: (1) better rejection of non-Gaussian interference, (2) superior rejection of noise, (3) availability of additional blind equalization algorithms, (4) more robust startup, (5) capability of rejecting CDMA interference, (6) better equalization of non-minimal phase channels, and (7) better compensation of non-linear distortion. On the negative side, with present neural net equalization techniques, there is no guarantee of reaching an optimal solution, and the convergence rate is very slow.

The ability of neural networks to reject interference can be viewed using different perspectives: 1) the neural net can create non-linear decision boundaries between signal states, 2) neural nets provide a means of implementing non-linear filters for rejecting non-Gaussian interference, and 3) neural nets can be used to identify specific error patterns.

3.2.1   Back-Propagation

Zengjun and Guangguo [94] (1992) describe a fractionally-spaced DF multilayer perceptron (FSDFMLP) for adaptive multilevel QAM digital mobile radio reception that can reject cochannel interference and AWGN simultaneously. The FSDFMLP is trained by a fast adaptive learning algorithm called the mixed gradient based fast learning algorithm with variable learning gain and selective updates (based on a combination of the steepest descent and the conjugate gradient methods). FSDFMLP can perform more efficiently than the conventional LMS based DF filter in the presence of multipath fading of channels with non-Gaussian interferences. Similarly, Zengjun and Guangguo [95] (1992) describe the complex neural-network-based adaptive DF filter (CNNDFF) for M-QAM digital communication reception systems. Experimental results indicate that the CNNDFF can simultaneously overcome the performance degradations due to multipath fading of channels and reject the non-Gaussian cochannel interferences efficiently. The convergence rate of the CNNDFF is significantly better than the standard back-propagation network.

Bijjani and Das [96] (1990) present a multilayer back-propagation perceptron model as a means of detecting a wideband signal in the presence of narrowband jammers and additive white noise. The non-linear neural network filter is shown to offer a faster convergence rate and overall better performance than the LMS Widrow-Hoff filter.

3.2.2   Radial Basis Function

Chen and Mulgrew [97] (1992) show results of applying the adaptive radial basis function (RBF) neural net to interference rejection and equalization. They show that an adaptive RBF neural net equalizer can implement the optimal Bayesian symbol-decision equalizer using a two-stage learning algorithm. The first stage is a supervised or decision-directed clustering algorithm which learns the centers of the desired signal states, and the second stage is a variation of an unsupervised k-means clustering algorithm for modeling the effect of the interference. In one example, the neural net provides an effective reduction in SINR by 7 dB over the transversal equalizer for a BER of 10^{-4}. The algorithm converges remarkably fast even compared to traditional equalization algorithms.

Cha and Kassam [98] (1992) also investigate the applicability of the RBF network in adaptive interference cancellation problems. An extended structure that combines a linear canceler with an RBF network is shown to be more robust than the structure using an RBF network only. A means for growing the RBF network in interference rejection applications is addressed by Howitt, et al. [99] (1994)

3.3 Non-Linear Techniques

By viewing noise cancellation as an input/output identification problem, Giannakis and Dandawate [100] (1990) develop designs using third-order statistics which are insensitive to corruption of the reference signal by additive Gaussian noise of unknown covariance. As a by-product of designing linear noise cancelers, a parametric time-delay estimate is readily available, and higher-order statistics can also be employed to design non-linear cancelers of the discrete Volterra-type, which maximize the output signal-to-noise ratio (SNR).

Valeev and Yazovskii [101] (1987) consider a method for construction of adaptive non-linear converters (ANC) as a preprocessor to a correlation receiver for improving immunity to non-Gaussian interference. The authors show how to construct the non-linearity to maximize output SNR.

Maulhardt, Davis, and May [102] (April 1986) present techniques for designing frequency-domain non-linear adaptive filters. These techniques make use of hierarchical memory structures that are trained to learn the appropriate transfer functions for a given signal and interference environment.

3.4 Spectral Correlation

An adaptive filter is a time-varying filter, where the filter coefficients change with time, minimizing some error criterion function. However, if the signal statistics change rapidly, a conventional adaptive filter is incapable of converging to the optimum solution. This is often the case in applications when the adaptive filter is used for filtering digitally modulated signals. These signals have periodic statistics and are generally referred to as cyclostationary signals which possess the property of spectral correlation. Spectral correlation is a measure of "spectral redundancy," and this property is the key to obtaining improved interference rejection capability over that of conventional time-independent filters.

Analog and digital carrier modulated signals, such as AM, digital QAM, PSK and FSK, exhibit correlation among spectral components separated by multiples of the keying rate and separated by the doubled carrier frequency plus multiples of the keying rate. Gardner and Venkataraman [103] (1990) observe that this spectral redundancy can be exploited to facilitate rejection of cochannel interference, while maintaining minimal signal distortion. Gardner and Brown [104] (1989) show how spectral redundancy is exploited by multichannel frequency shift filtering of the corrupted data and adding the results to implement a time-dependent filter. Gardner [105] (1993) develops some of the theoretical concepts underlying this type of filtering and summarizes the theory of optimal FREquency SHift (FRESH) filtering, which is a generalization of Wiener filtering, called cyclic Wiener filtering. He also shows how the performance depends on the signal's excess bandwidth.

A FRESH DF equalizer is a DF equalizer (DFE) where the forward filter is replaced by a bank of filters whose inputs are frequency-shifted. By exploiting the spectral redundancy of modulated signals, this technique improves the DFE performance in a cyclostationary environment. Hendessi, Hafez, and Sheikh [106] (1993) show that the performance of the FRESH-DFE is superior to that of a conventional DFE.

Mendoza, Reed, Hsia, and Agee [107] (1991) present two new blind adaptive filtering algorithms for interference rejection using time-dependent filtering structures that exploit cyclostationary signals. They show that the blind (i.e., operating without the use of an external training signal) time-dependent filtering algorithms can provide MSEs and BERs that are significantly lower than the MSEs and BERs provided using conventional time-independent adaptive filters (which are non-blind and training-sequence directed). The theoretical MSE of time-dependent filtering is also given by Reed and Hsia [108] (1990)

Greene, Reed, and Hsia [109] (1989) present the optimal time-dependent receiver (OTDR) and show it to be superior to the conventional matched filter receiver when cyclostationary interference is present. This performance advantage is explained by noting that the matched filter is periodic at the baud rate of the SOI; while the OTDR is periodic at the baud rate of the SOI and any other statistical periodicity of the received signal (including that of the interfering signal).

Nicolas and Lim [110] (1993) address the problem of transmitting digital HDTV signals in a co-channel interference (CCI) limited environment. They describe a new signal processing technique aimed at rejecting CCI from adjacent analog transmitters. The proposed scheme uses a form of joint DFE/trellis coded modulation to combat the interference. DFE can be used in the application by exploiting the cyclostationary properties of the interference. The technique has several advantages over methods previously proposed: 1) processing is constrained to the receiver, 2) the scheme is able to make use of powerful coding schemes, 3) the scheme is adaptive and 4) reception on conventional NTSC receivers is not affected by this scheme.

3.5 Novel Techniques

Shin and Nikias [111] (1993) introduce a new higher-order statistics-based adaptive interference canceler to eliminate additive narrowband and wideband interferences in environments where the interference is non-Gaussian and where a reference signal, which is highly correlated with the interference, is available. They demonstrate that the higher-order statistics-based adaptive algorithm performs more effectively than the second-order statistics-based adaptive algorithm not only for single and multiple narrowband interferences with/without Gaussian uncorrelated noise sources but also for wideband (AM and FM) interferences. Exploiting higher order statistics can lead to new blind adaptive filtering techniques for interference rejection.

Bar-Ness and Bunin [112] (1989) improve on a method for co-channel interference suppression and signal separation which uses the amplitude variation of the composite signal to estimate the parasitic phase modulation impinged on the strong desired signal by the weak interference signal. This estimate is then used to cancel out the distortion of the composite signal, revealing the desired signal. In the cancellation process, amplitude estimates for both signals were obtained from measurements. An adaptive method is proposed which improves these estimates and, hence, results in a better cancellation of interference. In comparison with nonadaptive methods, the adaptive approach exhibits an additional 21 dB of interference suppression.

4   CONCLUSION

Though finding their roots in military anti-jam research, interference rejection techniques are of increasing interest to industry because of their applicability to commercial wireless communications. Adaptive notch filtering is one of the earliest and simplest forms of adaptive interference rejection and has been supplemented by many new interference rejection techniques capable of rejecting interference with less distortion and under a wider variety of signal conditions. There remains much work, however, in determining the relative merits and practicality of the newer techniques.

5   REFERENCES

[1]   D. L. Schilling, G. R. Lomp, and J. Garodnick, "Broadband-CDMA overlay," *Proceedings of the 43rd IEEE Vehicular Technology Conference*, p.452-455, 1993.

[2]   D.J. Kennedy and E.K. Koh, "Frequency-reuse interference in TDMA/QPSK satellite systems," *Fifth International Conference on Digital Satellite Communications*, (Genoa, Italy), p. 99-107, March 23-26, 1981.

[3]   G. W. Travis and H. F. Lenzing, "Shipboard HF interference: problems and mitigation," *MILCOM '89: Bridging the Gap. Interoperability, Survivability, Security. Conference Record*, vol.1, p.106-10, October 1989.

[4]   R. Kohno, "Pseudo-noise sequences and interference cancellation techniques for spread spectrum systems-spread spectrum theory and techniques in Japan," *IEICE Transactions*, vol.E74, no.5, p.1083-92, May 1991.

[5]   L.B. Milstein, "Interference rejection techniques in spread spectrum communications," *Proceedings of the IEEE*, vol.76, no.6, June 1988.

[6]   G.J. Saulnier, "Suppression of narrowband jammers in a spread-spectrum receiver using transform-domain adaptive filtering," *IEEE Journal on Selected Areas in Communications*, vol.10, no.4, p.742-9, May 1992.

[7]   D.V. Compernolle and S. van Gerven, "Signal separation in a symmetric adaptive noise canceler by output decorrelation," *ICASSP-92: 1992 IEEE International Conference on Acoustics, Speech and Signal Processing*, vol.4, p.221-4, 23-26 March 1992.

[8]   J.F. Doherty, "Direct sequence spread spectrum interference rejecton using vector space projection techniques," *Virginia Tech's Third Symposium on Wireless Personal Communications. Proceedings*, p.9/1-9, June 1993.

48

[9]   J.F. Doherty, "A constrained LMS algorithm for interference rejection," *MILCOM '92 - 'Communications - Fusing Command, Control and Intelligence' Conference Record*, vol.2, p.696-700, 11-14 October 1992.

[10]   J.F. Doherty, "Linearly constrained interference rejection for improved spread spectrum performance," *SUPERCOMM/ICC '92. Discovering a New World of Communications*, vol.3, p.1257-61, 14-18 June 1992.

[11]   J.F. Doherty, "An adaptive technique for improving spread spectrum interference rejection," *Proceedings. RF Expo EAST*, p.385-7, 29-31 October 1991.

[12]   J. Gevargiz, P.K. Das, and L.B. Milstein, "Adaptive narrow-band interference rejection in a DS spread-spectrum intercept receiver using transform domain signal processing techniques," *IEEE Transactions on Communications*, vol.37, no.12, p.1359-66, December 1989.

[13]   J. Gevargiz, P.K. Das, and L.B. Milstein, "Implementation of a transform domain processing radiometer for DS spread spectrum signals with adaptive narrow-band interference exciser," *Proc. Int. Conf. Commun. (ICC'86)*, CH2314-3, 1986, p. 251-255.

[14]   J. Gevargiz, P.K. Das, and L.B. Milstein, "Performance of a transform domain processing DS intercept receiver in the presence of finite bandwidth interference," *GLOBECOM '86: IEEE Global Telecommunications Conference. Communications Broadening Technology Horizons. Conference Record*, vol.2, p.738-42, 1-4 Dec. 1986.

[15]   N.J. Bershad, "Error probabilities for DS spread-spectrum systems using an ALE for narrow-band interference rejection," *IEEE Transactions on Communications*, vol.36, no.5, p.588-95, May 1988.

[16]   Z.D. Stojanovic, M.L. Dukic, I.S. Stojanovic, "A new method for the narrow-band interference rejection in the direct sequence spread-spectrum systems using transversal filters," *Proceedings of MELECON '87: Mediterranean Electrotechnical Conference and 34th Congress on Electronics Joint Conference*, p.149-52, 24-26 March 1987.

[17]   Y. He, S.-F. Lei, P. Das, and G.J. Saulnier, "Suppression of narrowband jammers in a DS spread spectrum receiver using modified adaptive filtering technique," *GLOBECOM '88. IEEE Global Telecommunications Conference and Exhibition - Communications for the Information Age. Conference Record*, vol.1, p.540-5, 28 December 1988.

[18]   J.H. Lee and C.W. Lee, "Adaptive filters for suppressing irregular hostile jamming in direct sequence spread-spectrum system," *MILCOM 87: 1987 IEEE Military Communications Conference. 'Crisis Communications: The Promise and Reality'. Conference Record*, vol.1, p.118-22, 19-22 October 1987.

[19]   J.H. Lee and C.W. Lee, "A study on irregular narrow-band interference rejection in direct sequence spread-spectrum system using adaptive filter," *Proceedings of TENCON 87: 1987 IEEE Region 10 Conference 'Computers and Communications Technology Toward 2000'*, vol.3, p.1376-80, 25-28 August 1987.

[20]   G.J. Saulnier, K. Yum, and P. Das, "Narrow-band jammer suppression using an adaptive lattice filter," *Proceedings: ICASSP 87. 1987 International Conference on Acoustics, Speech, and Signal Processing*, vol.4, p.2113-16, 6-9 April 1987.

[21]   G.J. Saulnier, K. Yum, and P. Das, "The suppression of tone jammers using adaptive lattice filtering," *IEEE International Conference on Communications '87: Communications-Sound to Light. Proceedings*, vol.2, p.869-73, 7-10 June 1987.

[22]   J. Guilford and P. Das, "The use of the adaptive lattice filter for narrowband jammer rejection in DS spread spectrum systems," *IEEE International Conference on Communications 1985*, vol.2, p.822-6, 23-26 June 1985.

[23]   G.J. Saulnier, P.K. Das, and L.B. Milstein, "An adaptive digital suppression filter for direct-sequence spread-spectrum communications," *IEEE Journal on Selected Areas in Communications*, vol.SAC-3, no.5, p.676-86, September 1985.

[24]   G.J. Saulnier and P. Das, "Antijam spread spectrum receiver using LMS adaptive filtering techniques," *MILCOM '84. IEEE Military Communications Conference. Conference Record*, vol.3, p.482-7, 21-24 Oct. 1984.

[25]   G.J. Saulnier, P.K. Das, and L.B. Milstein, "Suppression of narrow-band interference in a direct sequence spread spectrum receiver in the absence of carrier synchronization," *1985 IEEE Military Communications Conference: MILCOM '85. Conference Record*, vol.1, p.13-17, 20-23 October 1985.

[26]   P.W. Bandy and D.J. Krause, "Elimination of narrowband noise in spread-spectrum systems," *Conference Proceedings. 28th Midwest Symposium on Circuits and Systems*, p.784-7, 19-20 August 1985.

[27]   P.W. Baier and K.J. Friederichs, "A nonlinear device to suppress strong interfering signals with arbitrary angle modulation in spread spectrum receivers." *IEEE Trans. Commun.*, vol. COM-33, p. 300-302, Mar. 1985.

[28]   R.L. Fox, S.J. MacMullen, and A. Nauda, "Comparison of interference reduction techniques in the demodulation of bauded angle-modulated signals," *Proceedings of the 1985 Summer Computer Simulation Conference*, p.259-64, 22-24 July 1985.

[29]   A. Mammela, "The performance of adaptive interference suppression filters used in PN spread-spectrum systems," *EUROCON 88: 8th European Conference on Electrotechnics. Conference Proceedings on Area Communication*, p.126-9, 13-17 June 1988.

[30]   C.S. Lindquist, K.C. Severance, and H.B. Rozran, "Frequency domain algorithms for adaptive line enhancer and adaptive noise canceler systems," *Conference Record Eighteenth Asilomar Conference on Circuits, Systems and Computers*, p.233-7, 5-7 Nov. 1984.

[31]   E. Plotkin and D. Wulich, "A comparative study of two adaptive algorithms for suppression of a narrow-band interference," *Journal: Signal Processing*, vol.4, no.1, p.35-44, January 1982.

[32]   E. Plotkin and A. Plotkin, "An adaptive approach to suppress powerful impulsive interference," *Journal: Signal Processing*, vol.4, no.1, p.25-33, January 1982.

[33]   J.W. Ketchum and J.G. Proakis, "Adaptive algorithms for estimating and suppressing narrow-band interference in PN spread-spectrum systems," *IEEE Transactions on Communications*, vol.30, no.5, pt.1, p. 913-24, May 1982.

[34]   L. Li and L.B. Milstein, "The use of adaptive filters for narrowband interference rejection," *NTC '81. IEEE 1981 National Telecommunications Conference. Innovative Telecommunications - Key to the Future*, vol.1, p. B8.4/1-4, December 1981.

[35]   A.J. Dunlop and I.R. Whitworth, "Adaptive spread spectrum modulation," *Conference on Communications Equipment and Systems*, p.183-6, 16-18 April 1980.

[36]   C.N. Pateros and G.J. Saulnier, "Adaptive correlator receiver performance iier, "Adaptive correlator receiver performa spectrum communication," *MILCOM '92 - 'Communications - Fusing Command, Control and Intelligence' Conference Record*, vol.2, p.427-31, 11-14 October 1992.

[37]   C.N. Pateros and G.J. Saulnier, "Interference suppression and multipath mitigation using an adaptive correlator direct sequence spread spectrum receiver," *SUPERCOMM/ICC '92. Discovering a New World of Communications*, vol.2, p.662-6, 14-18 June 1992.

[38]   T. Ogawa, I. Sasase, and S. Mori, "Suppression of narrow-band interference and multipath by spread-spectrum receiver using decision-feedback filters." *IEEE Pacific Rim Conference on Communications, Computers and Signal Processing*, vol.2, p.673-6, 9-10 May 1991.

[39]   M. Miyagi, T. Ogawa, I. Sasase, and S. Mori, "Suppression of CW interference and filtered noise in QPRS systems using decision-feedback filters," *ICASSP 90. 1990 International Conference on Acoustics, Speech and Signal Processing*, vol.3, p.1703-6, 3-6 April 1990.

[40]   T. Ogawa, I. Sasase, and S. Mori, "Suppression of CW interference and colored noise in QPSK system using decision-feedback filters," *Transactions of the Institute of Electronics, Information and Communication Engineers E*, vol.E72, no.7, p.804-10, July 1989.

[41]   M.L. Dukic, Z.D. Stojanovic, and I.S. Stojanovic, "Performance of direct-sequence spread-spectrum receiver using decision feedback and transversal filters for combating narrowband interference," *IEEE Journal on Selected Areas in Communications*, vol.8, no.5, p.907-14, June 1990.

[42]   M.L. Dukic, Z.D. Stojanovic, I.S. Stojanovic. "A new direct sequence spread spectrum receiver using decision feedback and transversal filters for rejection of the narrow-band interference and errors caused by signal distortion," *MELECON '89: Mediterranean Electrotechnical Conference Proceedings. Integrating*

Research, Industry and Education in Energy and Communication Engineering. p.395-8, 11-13 April 1989.

[43] Z.D. Stojanovic, M.L. Dukic, and I.S. Stojanovic. *Proceedings fo MELECON '87: Mediterranean Electrotechnical Conference and 34th COngress on Electronics Joint Conference.* March 1987.

[44] M.L. Dukic, D.O. Cuberovic, Z.D. Stojanovic, and I.S. Stojanovic. "Performance analysis of DS spread-spectrum receiver using decision feedback and transversal interference suppression filters under multiple narrow-band interference," *Communication Systems: Towards Global Integration. Singapore ICCS '90. Conference Proceedings,* vol.2, p. 25-2/1-5, 5-9 Nov. 1990

[45] B. Shah and G.J. Saulnier, "Adaptive jammer suppression using decision feedback in a spread-spectrum receiver," *MILCOM 88. 21st Century Military Communications -What's Possible? Conference Record. 1988 IEEE Military Communications Conference,* vol.3, p.989-95, 23-26 October 1988.

[46] F. Liu and L. Li, "Rejection of finite-bandwidth interference in QPSK system using decision-feedback filters," *IEEE International Conference on Communications '87: Communications-Sound to Light. Proceedings,* vol.2, p.879-83, 7-10 June 1987.

[47] F. Takawira and L.B. Milstein, "Narrowband interference rejection in PN spread spectrum communications systems using decision feedback filters," *MILCOM '86,* vol.2, p.20.4/1-5, October 1986.

[48] L. Li and L.B. Milstein, "Rejection of CW Interference in QPSK systems using decision-feedback filters," *IEEE Transactions on Communications,* vol. COM-31, no.4, April 1983.

[49] W. Libing, B. Guangguo, and W. Boxiu, "Suppression of FM interference in QAM systems using adaptive decision-feedback filters," *China 1991 International Conference on Circuits and Systems. Conference Proceedings,* vol.1, p.161-3, 16-17 June 1991.

[50] A.M.J. Goiser and M.K. Sust, "Adaptive interference rejection for non-coherent digital direct sequence spread spectrum receivers," *GLOBECOM '90: IEEE Global Telecommunications Conference and Exhibition. 'Communications: Connecting the Future',* vol.1, p.285-90, 2-5 Dec. 1990.

[51] A.M.J. Goiser and M.K. Sust, "Adaptive interference rejection in a digital direct sequence spread spectrum receiver," *IEEE Military Communications Conference - MILCOM '89.* Part 2, p. 514-520, 1989

[52] F. Amoroso, "Performance of the Adaptive A/D Converter in Combined CW and Gaussian Interference," *MILCOM,* 1984 IEEE.

[53] F. Amoroso, "Adaptive A/D converter to suppress CW interference in DSPN spread spectrum communications," *IEEE Trans. Commun.,* vol. COM-31, p. 1117-1123, October 1983.

[54] F. Amoroso and J.L. Bricker, "Performance of the adaptive A/D converter in combined CW and Gaussian interference," *IEEE Trans. Commun.,* vol. COM-34, p. 209-213, March 1986.

[55] F.J. Pergal, "Adaptive threshold A/D conversion techniques for interference rejection in DSPN receiver applications," in *IEEE Military Communications Conf.,* p. 4.7.1-4.7.7, October 1987.

[56] M.B. Pursley, "The role of spread-spectrum in packet radio networks," *Proceedings of the IEEE,* vol.75, no.1, p.116-134, January 1987.

[57] S. Verdu, "Minimum probability of error for asynchronous multiple-access channel," *IEEE Trans. on Info. Theory,* vol. IT-32, no. 5, p. 642-651, September 1986.

[58] R. Lupas and S. Verdu, "Linear multiuser detectors for synchronous code-division multiple-access channel," *IEEE Trans. Commun.,* vol.IT-32, no.1, p.85-96, January 1986.

[59] B. Aazhang, B.-P. Paris, and G.C. Orsak, "Neural networks for multi-user detection in code-division multiple-access communications," *IEEE Transactions on Communications,* vol. 40, no. 7, p. 1212-1222, July 1992.

[60] P.N. Monogioudis, R. Tafazolli, and B.G. Evans, "Linear adaptive fractionally spaced equalization of CDMA multiple-access interference," *Electronics Letters,* vol.29, no.21, p.1823-5, 14 Oct. 1993.

[61] M.E. Davis and L.B. Milstein, "Anti-jamming properties of a DS-CDMA equalization filter," *Proceedings of the 12th Annual IEEE Military Communications Conference,* vol.3, p 1008-1012, 1993.

[62] P.B. Rapajic and B.S. Vucetic, "Linear adaptive fractionally spaced single user receiver for asynchronous CDMA systems," *Proceedings of the 1993 IEEE International Symposium on Information Theory,* p 45, 1993.

[63]   M. Abjulrahman, D.D. Falconer, and A.U.H. Sheikh, "Equalization for Interference Cancellation in Spread Spectrum Multiple Access Systems," *Proceedings VTC*, May 1992.

[64]   U. Madhow and M. Honig, "Error probability and near-far resistance of minimum mean squared error interference suppression schemes for CDMA," *GLOBECOM '92. Communication for Global Users. IEEE Global Telecommunications Conference. Conference Record*, vol.3, p.1339-43, 6-9 December 1992.

[65]   U. Madhow and M.L. Honig, "Minimum mean squared error interference suppression for direct-sequence spread-spectrum code-division multiple-access," *ICUPC '92 Proceedings*, p.10.04/1-5, 29 Sept.-1 Oct. 1992.

[66]   R.D. Holley and J.H. Reed, "Time dependent adaptive filters for interference cancellation in CDMA systems," unpublished master's thesis, Mobile & Portable Radio Research Group (MPRG-TR-93-15A), Bradley Department of Electrical Engineering, Virginia Polytechnic Institute and State University, October 1993.

[67]   V. Aue and J.H. Reed, "CDMA demodulation and interference rejection using an optimal time-dependent filter," unpublished master's thesis, Mobile & Portable Radio Research Group, Bradley Department of Electrical Engineering, Virginia Polytechnic Institute and State University, February 1994.

[68]   M.K. Varansi and B. Aazhang, "Multi-stage detection in asynchronous code-division multiple-access communications," *IEEE Trans. on Comm.*, vol. COM-38, no. 4, p. 509-519, April 1990.

[69]   V. Bais, "Improved receiver design for CDMA systems." MPRG Technical Report, Virginia Polytechnic Institute and State University, June 1993.

[70]   M. Kawabe, T. Kato, A. Kawahashi, T. Sato, and A. Fukasawa, "Advanced CDMA scheme based on interference cancellation," *IEEE Vehicular Technology Conference 1993*, p.448-451, 1993.

[71]   S.S.H. Wijayasuriya, J.P. McGeehan, and G.H. Norton, "RAKE decorrelation as an alternative to rapid power control in DS-CDMA mobile radio," *IEEE Vehicular Technology Conference* (VTC), Piscataway, NJ, May 1993.

[72]   S. Striglis, A. Kaul, N. Yang, and B. Woerner, "A multi-stage RAKE receiver for improved CDMA performance," *IEEE Vehicular Technology Conference* (VTC), May 1994.

[73]   P. Patel and J.M. Holtzman, "Analysis of DS /CDMA successive interference cancellation scheme using correlation," *GLOBECOM '93*, December 1993.

[74]   R. Kohno, H. Imai, M. Hatori, and S. Pasupathy, "An adaptive canceller of cochannel interference for spread-spectrum multiple-access communication networks in a power line," *IEEE Journal on Selected Areas in Communications*, vol. 8, no. 4, May 1990.

[75]   R.S. Mowbray, R.D. Pringle, and P.M. Grant, "Increased CDMA system capacity through adaptive cochannel interference regeneration and cancellation," *IEE Proceedings I (Communications, Speech and Vision)*, vol.139, no.5, p.515-24, October 1992.

[76]   R.S. Mowbray, R.D. Pringle, and P.M. Grant, "Adaptive CDMA cochannel interference cancellation," *Signal Processing VI - Theories and Applications. Proceedings of EUSIPCO-92, Sixth European Signal Processing Conference*, vol. 3, p.1591-4, 24-27 August 1992.

[77]   S.G. Glisic and M.D. Pajkovic, "Rejection of FHMA signal in DS spread spectrum system using complex adaptive filtering," *Military Communications in a Changing World MILCOM, 91. Conference Record*, vol.1, p.365-9, 4-7 Nov. 1991.

[78]   S.G. Glisic and M.D. Pajkovic, "Rejection of FHMA signal in DS spread spectrum system using complex adaptive filtering," *MILCOM 90. A New Era. 1990 IEEE Military Communications Conference. Conference Record*, vol.1, p.349-53, October 1990.

[79]   R.A. Iltis, "Interference cancellation for enhanced detection of frequency-hopped signals," *ICASSP 86 (Toyko)*, vol.2, p. 973-976, April 1986.

[80]   J.H. Reed and B. Agee, "A maximum-likelihood approach to detecting frequency hopping spread spectrum signals," *1992 Asilomar Conference on Signals, Systems, and Computers*, 1992.

[81]   F.A. Bishop and R.S. Leahy, "Enhancement of frequency hopped signals by convergence bandwidth discrimination," *1985 IEEE Military Communications Conference: MILCOM '85. Conference Record*, vol.2, p.334-8, 20-23 Oct. 1985.

52

[82] T.A. Gulliver, "Order Statistics Diversity Combining in Worst Case Noise and Multitone Jamming," *MILCOM 91*, vol.1, p. 385-389, November 1991.

[83] L.M. Garth and H.V. Poor, "Narrowband interference suppression in impulsive channels," *IEEE Transactions on Aerospace and Electronic Systems*, vol.28, no.1, p.15-34, January 1992.

[84] L. Garth, R. Vijayan, and H.V. Poor, "A new approach to interference suppression in spread-spectrum systems," *Military Communications in a Changing World MILCOM, 91. Conference Record*, vol.1, p.375-9, 4-7 November. 1991.

[85] R. Vijayan and H.V. Poor, "Nonlinear techniques for interference suppression in spread-spectrum systems," *IEEE Transactions on Communications*, vol.38, no.7, p.1060-5, July 1990.

[86] F. Dominique and F. Joseph, "Frequency domain median filtering for narrow band interference rejection in BPSK-PN signals," *IEEE TENCON '93*, 1993.

[87] J.H. Higbie, "Adaptive nonlinear suppression of interference," *MILCOM 88: 21st Century Military Communications - What's Possible?*, (San Diego, CA), p 381-389, 1988.

[88] O.W. Kwon, C.K. Un, and J.C. Lee, "Performance of constant modulus adaptive digital filters for interference cancellation," *Signal Processing*, vol.26, no.2, p.185-96, February 1992.

[89] M.J. Rude and L.J. Griffiths, "An untrained, fractionally-spaced equalizer for co-channel interference environments," *Twenty-Fourth Asilomar Conference on Signals, Systems and Computers*, vol.1, p.468-72, 5-7 November 1990.

[90] E.R. Ferrara, Jr., "A method for canceling interference from a constant envelope signal," *IEEE Transactions on Acoustics, Speech and Signal Processing*, vol.ASSP-33, no.1, p.316-19, February 1985.

[91] R.P. Gooch and B. Daellenbach, "Prevention of interference capture in a blind (CMA-based) adaptive receive filter," *Twenty-Third Asilomar Conference on Signals, Systems ands Computers*, vol.2, p.898-902, November 1989.

[92] E.H. Satorius, S. Krishnan, X. Yu, L.J. Griffiths, I.S. Reed, and T. Truong, "Suppression of narrowband interference via single channel adaptive preprocessing," *Twenty-Second Asilomar Conference on Signals, Systems and Computers*, vol.1, p.270-3, November 1988.

[93] I. Howitt, J.H. Reed, V. Vemuri, and T.C. Hsia, "Recent Developments in Applying Neural Nets to Equalization and Interference Rejection," *Virginia Tech's Third Symposium on Wireless Personal Communications. Proceedings*, June 1993.

[94] Zengjun Xiang and Guangguo Bi, "Fractionally spaced decision feedback multilayer perceptron for adaptive MQAM digital mobile radio reception," *SUPERCOMM/ICC '92. Discovering a New World of Communications*, vol.3, p.1262-6, 14-18 June 1992.

[95] Zengjun Xiang and Guangguo Bi, "Complex neuron model with its applications to M-QAM data communications in the presence of co-channel interferences," *ICASSP-92: 1992 IEEE International Conference on Acoustics, Speech and Signal Processing*, vol. 2, p.305-8, 23-26 March 1992.

[96] R. Bijjani and P.K. Das, "Rejection of narrowband interference in PN spread-spectrum systems using neural networks," *GLOBECOM '90: IEEE Global Telecommunications Conference and Exhibition. 'Communications: Connecting the Future'*, vol.2, p.1037-41, 2-5 Dec. 1990.

[97] S. Chen and B. Mulgrew, "Overcoming co-channel interference using an adaptive radial basis function equalizer," *Signal Processing*, vol. 28, p. 91-107, 1992.

[98] I. Cha and S.A. Kassam, "Interference cancellation using radial basis function networks," *IEEE Sixth SP Workshop on Statistical Signal and Array Processing Conference Proceedings*, p.221-4, 7-9 Oct. 1992.

[99] I. Howitt, J.H. Reed, R. Vemuri, T.C. Hsia, "RBF growing algorithm applied to the equalization and co-channel interference rejection problem," *IEEE World Congress on Computational Intelligence /International Conference on Neural Networks*, June 26 - July 2, 1994, Orlando, Florida.

[100] G.B. Giannakis and A.V. Dandawate, "Linear and non-linear adaptive noise cancelers," *ICASSP 90. 1990 International Conference on Acoustics, Speech and Signal Processing*, vol.3, p.1373-6, 3-6 April 1990.

[101] V.G. Valeev and A.A. Yazovskii, "Adaptive nonlinear converters for suppression of non-Gaussian interference," *Radioelectronics and Communication Systems*, vol.30, no.8, p.60-3, 1987.

[102]   M. Maulhardt, A.M. Davis, and J. May, "Numerical design of nonlinear adaptive filters," *ICASSP 86 Proceedings. IEEE-IECEJ-ASJ International Conference on Acoustics, Speech and Signal Processing*, vol.3, p.2131-4, 7-11 April 1986.

[103]   W.A. Gardner and S. Venkataraman, "Performance of optimum and adaptive frequency-shift filters for cochannel interference and fading," *Conference Record. Twenty-Fourth Asilomar Conference on Signals, Systems and Computers*, vol.1, p.242-7, 5-7 Nov. 1990.

[104]   W.A. Gardner and W.A. Brown, "Frequency-shift filtering theory for adaptive co-channel interference removal," *Twenty-Third Asilomar Conference on Signals, Systems ands Computers*, vol.2, p.562-7, November 1989.

[105]   W.A. Garner, "Cyclic Wiener filtering: theory and method," *IEEE Transactions on Communications*, vol. 41, no. 1, p. 151-163, January, 1993.

[106]   F. Hendessi, H.M. Hafez, and A.U.H. Sheikh, "Structure and performance of FRESH-decision feedback equalizer in the presence of adjacent channel interference," *Proceedings of the 43rd IEEE Vehicular Technology Conference*, p 641-644, 1993.

[107]   R. Mendoza, J.H. Reed, T.C. Hsia, and B.G. Agee, "Interference rejection using the time-dependent constant modulus algorithm (CMA) and the hybrid CMA/spectral correlation discriminator," *IEEE Transactions on Signal Processing*, vol. 39, no. 9, September 1991.

[108]   J. H. Reed and T. C. Hsia, "The performance of time-dependent adaptive filters for interference rejection," *IEEE Transactions on Acoustics, Speech, and Signal Processing*, vol. 38, no. 8, August 1990.

[109]   C. D. Greene, J. H. Reed, and T. C Hsia, "An optimal receiver using a time -dependent adaptive filter," *MILCOM 1989*, 1989.

[110]   J. J. Nicolas and J. S. Lim, "Equalization and interference rejection for the terrestrial broadcast of digital HDTV," Research Laboratory of Electronics, MIT, 1993.

[111]   D.C. Shin and C.L. Nikias, "Adaptive noise canceler for narrowband/wideband interferences using higher-order statistics," *1993 IEEE International Conference on Acoustics, Speech and Signal Processing*, v 3, p. 111.364-366, 1993.

[112]   Y. Bar-Ness and B.H. Yeheskel; Bunin, "Adaptive co-channel interference cancellation and signal separation method," *IEEE International Conference on Communications - ICC'89*, p 825-830, 11-14 June 1989

3

Spread Spectrum for Wireless Communications

R. M. Buehrer and B. D. Woerner

Mobile and Portable Radio Research Group
Bradley Department of Electrical Engineering
Virginia Polytechnic Institute and State University
Blacksburg, VA 24061-0111

Abstract

Though originally developed as a military technology, spread-spectrum communications has achieved widespread commercial acceptance in the last several years. The increased interest in spread spectrum for wireless communications can be attributed to the high spectral efficiency of Code Division Multiple Access techniques, and to the natural resistance of spread spectrum signals to the effects of multipath propagation. This paper introduces the basic concepts of spread spectrum communication, discusses current analytic models for spread spectrum systems and discusses important current applications of spread spectrum technology to wireless systems. The paper also highlights current research issues in spread spectrum communications, providing both a tutorial introduction to the subject and a survey of current work.

1. Introduction

The interest in wireless communications has grown dramatically in the last five years. This growth has been seen in several forms of wireless including cellular telephony, radiopaging, personal communication systems, and even indoor applications such as wireless local area networks and factory implementations.

This increased interest in wireless has brought more focus on the particular problems of tetherless communications. These problems include propagation effects, in particular multipath propagation, capacity limits due to spectrum availability, and the need for asynchronous access.

One possible method of mitigating the aforementioned problems is the use of spread spectrum communications. Spread spectrum promises several benefits. The two benefits of most concern to the wireless communications engineer are its promise of higher capacity and its ability to resist multipath interference. Other benefits include resistance to jamming, a low probability of intercept, a soft capacity limit, and the provision of asynchronous access.

The following paper describes the basics of spread spectrum, derives performance bounds for such systems, details issues of importance when designing spread spectrum systems, and outlines some current applications. The fundamental principles of spread spectrum are provided in section two. Section three outlines its performance in the presence of multiple access interference as well as multipath distortion. The fourth section outlines current applications of spread spectrum technology such as cellular telephony and position location. Section five describes research issues, and conclusions are presented in the final section.

2. Spread Spectrum Fundamentals

The term 'spread spectrum' refers to any system which satisfies three basic conditions [1]. The first condition is that the carrier signal occupies a bandwidth much larger than $1/T_s$ where T_s is the message symbol duration. One familiar example of this is simple frequency modulation. Since the message signal is used to modulate the frequency, an FM signal occupies a much larger bandwidth than is necessary. Secondly, spread spectrum refers to not just the occupation of excessive bandwidth, but to a signal which in addition has pseudorandom properties. These signals are pseudorandom inasmuch as they appear to be unpredictable. They are of course not truly random and can be reproduced by deterministic means. The third condition for spread spectrum is that reception of the signal is accomplished via cross correlation with a locally generated version of the pseudorandom carrier.

There are two primary ways of generating such pseudorandom signals. Direct sequence spread spectrum (DS/SS) is currently more commercially important and is the focus of this paper. DS/SS involves multiplying the outgoing symbol stream by another symbol sequence with a much smaller symbol duration. These smaller symbols are referred to as 'chips'. This in turn increases the overall signaling rate (but not the data symbol rate) as depicted in Figure 1. This increases the occupied bandwidth. The spreading sequence is a known pseudorandom sequence with each element existing on {+1, -1}. This code is also known as a signature sequence since each user has a different spreading code. The second method for generating spread spectrum signals is called frequency hopping. As the name implies, frequency hopping (FH) involves hopping the carrier frequency of the outgoing signal in a prescribed fashion. This again greatly increases the occupied bandwidth. Hybrid formats of the two schemes are also used, as well as a third format called 'chirp' modulation. DS/SS is the more commercially important and will be the focus of attention here.

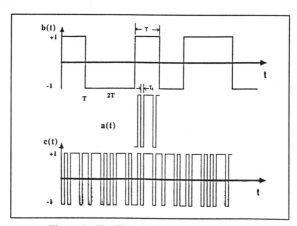

Figure 1: The Time-Domain Waveform

The original reason for spreading the signal over more bandwidth than necessary was two fold. Because spread spectrum spreads the signal energy over an extremely wide bandwidth, it hides the signal from other receivers within the noise floor. In order to receive the transmission, the receiver must correlate the signal with a locally generated version of the pseudorandom carrier, thus making the signal resistant to detection and interception by an unauthorized user. The second reason is that the signal is resistant to jamming since the jammer would need to produce enough energy to overcome the processing gain. That is, when the signal is correlated with the pseudorandom carrier, the desired signal is despread to its original bandwidth. The factor by which this bandwidth is decreased (and subsequently by which the power level is increased) is called the processing gain, N. In this process a jamming signal which is uncorrelated to the spreading sequence, is spread by a factor N. This reduction in the jamming power means that the jammer must increase his power by N to achieve the original desired interference level. Understandably, spread spectrum found its original applications in the military [1].

Currently, spread spectrum is now finding widespread civilian application (e.g. Qualcomm's IS-95 standard) [2]. There are two principal reasons for the use of spread spectrum in civilian systems: (1) greater capacity [3] and (2) resistance to multipath fading [4]. If proper codes are developed for each separate user, that is codes which have low cross-correlation and auto-correlation properties, multiple users can occupy the same spectrum. In other words, by choosing the codes properly we can create several orthogonal carriers. This application of spread spectrum is referred to as Code Division Multiple Access (or CDMA).

In reality, not all signals in a system will be perfectly orthogonal. Additionally, due to different delays experienced by different signals in route to a particular receiver, all sequences will not be synchronous with a particular user's signature sequence. Thus, there will be some correlation between the shifted versions of interfering signals and the desired user's signature sequence. This is of course multiple access or co-channel interference. It has been shown that if the chip sequence is large (>30) and the correlation properties are sufficiently low, the central limit theorem implies that this multiple access interference can be modeled as Gaussian noise [5] as will be discussed in section Three.

An additional implication of this property of low cross correlation is that, unlike TDMA or FDMA, there is no hard limit to the number of users allowed in the system at one time. Users are allowed to enter the system randomly, and there are not blocked calls per se. As the number of users increases, the signal quality of all users degrades gracefully, placing a soft limit on the number of users [6].

A main motivation for using CDMA in mobile radio is its resistance to multipath fading. At the receiver r(t) is demodulated (multiplied by the carrier frequency) and correlated with the signature sequence. Ideally, the signature sequence is completely random and thus has no correlation with a shifted version of itself. If the excess delay is greater than one chip duration, the correlation with each of the multipath components is then zero. Thus at the output of the correlator we see only the correlation of the transmitted data sequence with itself which is equal to $\sqrt{2P}$ and the correlation with AWGN. Effectively, we have eliminated multipath. Of course in practice the correlation is not identically zero, and thus we still

58

get some intersymbol interference. However, as we have shown, CDMA greatly reduces this interference. Again, as in the case of multiple access interference, this low amount of correlation can be modeled as AWGN for large chip sequences.

A number of excellent tutorial references are available on the theory of spread-spectrum communications [1,7,8]. Although most textbook treatments of spread spectrum emphasize the antijam and low probability of intercept applications of spread spectrum over the commercial applications, the analysis is readily adaptable to CDMA and antimultipath applications, and there are now many widely available papers which treat these applications [3,9]. The book [10] is also an enormously useful reference on practical hardware implementation of spread-spectrum systems.

3. Performance of Spread-Spectrum

In this section, we discuss the performance of DS/SS systems under a variety of conditions of interest to wireless system designers, including multiple access interference, multipath fading, near/far conditions and finally the performance of CDMA systems in a cellular environment. Over the last decade, considerable work has focused on the development of techniques to accurately predict the performance of DS/SS systems in these and similar environments. Performance evaluation techniques can be broadly classified into two categories: analytic and simulation. However, there has been a convergence between these areas. Analytic techniques are increasingly incorporating techniques to cope with complex systems designs and channel models, and Monte Carlo simulations are increasingly relying on semi-analytic methods such as importance sampling.

3.1 Multiple Access Performance

A DS/SS with K multiple access users is depicted in Figure 2. The k th user transmits the signal

$$s_k(t) = \sqrt{2P} a_k(t) b_k(t) \cos(\omega_c t + \phi_k), \qquad (1)$$

Where $a_k(t)$ is the signature signal of the k th user, $b_k(t)$ is the data signal of the k th user, and the signal is modulated by a carrier waveform with frequency ω_c, power P, and phase ϕ_k.

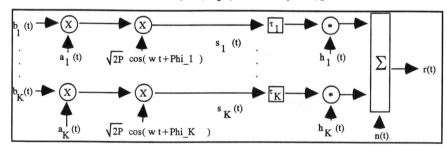

Figure 2: A Generalized CDMA System Model

The signal model can be generalized in a number of ways. Throughout this paper, we will assume that the chip waveform $\psi(t) = p_{T_c}(t)$ is a rectangular pulse of duration T_c, but the analysis can be generalized to arbitrary pulse shapes. The choice of pulse shape does not significantly affect performance although it does effect spectral usage. For example, the IS-95 cellular system employs overlapping chip waveforms for a sharp spectral cutoff [2]. (This standard will be discussed in more detail in section 4.) The model can be generalized to nonbinary signaling by selecting $a_k(t) \in \mathbf{A}^{(k)}$ from a nonbinary set of allowable sequences $\mathbf{A}^{(k)}$. The Walsh functions employed in the IS-95 standard are selected from such a set. The ratio $N = T/T_c$ is the processing gain of the system. Many analytic works have assumed that the code repeat rate is identical to the processing gain of the systems, but this is not necessarily the case. Analytic techniques can easily be generalized to the case of nonrepeating signature sequences. The IS-95 system employs a long signature sequences which is for all practical purposes infinite in length. Provided that the cross-correlation properties do not change significantly over time, it is the processing gain and not the sequence length which determines multiple access capability. Long sequence lengths have the advantage of improved security and availability of sequences for a large user community. However, systems employing long sequences are not able to exploit cyclostationary properties for fast synchronization and interference rejection.

The received signal r(t) is the sum of K delayed and distorted transmitted signals. Signals are delayed by a uniformly distributed delay τ_k and convolved with the channel impulse response $h_k(t)$. Of course $n(t)$ is a Gaussian noise process with two-sided power spectral density $N_o/2$. As mentioned earlier, reception is accomplished by correlating the received signal with the appropriate signature sequence to produce a decision variable. The decision variable for the i th transmitted bit of user 1 is

$$z_i^{(1)} = \int_{(i-1)T + \tau_1}^{iT + \tau_1} r(t)a(t-\tau_1)\cos[\omega_c(t-\tau_1) + \phi_1]dt . \tag{2}$$

If $b_{1,i} = +$, then the bit will be received in error provided that $z_i^{(1)} < 0$.

We wish to calculate the Bit Error Rate (BER) of a CDMA system as $P_b = Pr\left[Z_i^{(1)} < 0 | b_{1,i} = +1\right]$. Because $r(t)$ is a linear combination of signals, we may rewrite (2) as

$$Z_i^{(1)} = I_1 + \sum_{k=2}^{K} I_k + \xi , \tag{3}$$

where

$$I_1 = \int_0^T s_1(t)a_1(t)\cos(\omega_c t)dt = \sqrt{\frac{P}{2}}T \tag{4}$$

is the response of the receiver to the desired signal from user 1,

$$\xi = \int_0^T n(t)a_1(t)\cos(\omega_c t)dt$$

is a Gaussian random variable representing noise with mean zero and variance

$$E[\xi^2] = \frac{N_o T}{4} ,$$

and

$$I_k = \int_0^T s_k(t - \tau_k) a_1(t) \cos(\omega_c t) dt$$

(5)

represents the multiple access interference from user k. The difficulty lies in evaluating the distribution of multiple access interference. We may rewrite the expression for interference as

$$I_k = \sqrt{\frac{P}{2}} \cos(\theta_k) \left[b_{k,-1} \int_0^{\tau_k} a_k(t - \tau_k) a_1(t) dt + b_{k,0} \int_{\tau_k}^T a_k(t - \tau_k) a_1(t) dt \right],$$

(6)

which makes explicit the dependence of interference on the cross-correlation of the sequences between the two users. Ideally, we would like $\int a_k(t - \tau_k) a_1(t) dt \approx 0$ for any value of τ_k. This is not possible in practice, but it is possible to find many sets of sequences with good cross-correlation properties [11].

3.2 The Gaussian Approximation

One frequent approximation is to assume that I_k is composed of the cumulative effects of N random chips from interferer k. The Central Limit Theorem implies that the sum of these tiny effects will tend towards a Gaussian distribution. Since there are $K-1$ independent, identically distributed interferers, the total multiple access interference $I = \sum_{k=2}^K I_k$ may be approximated by a Gaussian random variable. The Gaussian approximation, a convenient analytic tool which is employed frequently [5,12], results in the expression

$$P_b = Q \left\{ \frac{1}{\sqrt{\frac{(K-1)}{3N} + \frac{N_o}{2E_b}}} \right\},$$

(7)

where $Q(.)$ is the well-known Q-function. When K=1, (7) reduces to the expression for error probability of a BPSK channel. As $\frac{E_b}{N_o} \to \infty$, $P_b = Q\left\{ \sqrt{\frac{3N}{K-1}} \right\}$ indicating the presence of an irreducible error floor for the multiple access interference limited case.

3.3 Other Analytic Expressions for Error Probability

Despite its convenience, the Gaussian approximation has limitations when used to predict system performance. While, the Central Limit Theorem implies that I will tend towards a Gaussian distribution near the center of it's distribution, convergence may be slow at the tails. Unfortunately, it is the extreme values of I which tend to produce errors. As a result, the Gaussian approximation is optimistic for low BER [13]. The problem is worsened in situations where there is a single strong interferer or a single strong multipath component. This is of concern, because low BER's are difficult to simulate, so we particularly desire analytic techniques which are accurate at low BER's.

Analytic techniques have been developed to accurately model the effects of multiple access interference. Rather than approximate the I_k as a Gaussian random variable, Lehnert and has developed a method to model the true probability density function of the multiple access interference $p_{I_k}(x)$ [13]. Once $p_{I_k}(x)$ is known, numerical integration may be used to develop arbitrarily tight bounds on $P_b = \Pr[Z_i^{(1)} < 0 | b_{1,i} = +1]$. This computation is numerically intensive, and several approximations are available. Morrow and Lehnert have proposed an improved Gaussian approximation [14] in which the variance is first evaluated conditionally, and Holtzman has developed a series approximation technique [15], both of which agree with exact results under a wide range of conditions.

3.4 The Near/Far Problem

Equation (7) has been derived under the assumption that signals from all users are received with the same signal power. Within a cellular system, such an assumption may be reasonable for the forward link in which all transmissions from a base station originate with the same signal power and follow the same transmission path; however, this assumption is not realistic on the reverse link from the mobile to the base station because signals originate from diverse and moving locations. This effect can be partially compensated for by power control which takes two forms. Under open loop power control a mobile adjusts its power based on the strength of the signal it receives from the base station. This presumes that a reciprocal relationship exists between the forward and reverse links, an assumption which may not hold if the links operate in different frequency bands. As a result, closed loop power control is often required in which the base station orders changes in the mobile's transmitted power.

The effects of the near/far problem and power control may be investigated using the analytic techniques described above with non-equal signal powers. The accuracy of the Gaussian approximation is reduced in the case of the near/far problem, but not entirely eliminated because the Central Limit Theorem applies to the processing gain N as well as the number of interferers K. Exact analysis techniques are still valid. As an example, the effectiveness of power control in the IS-95 system is investigated in [16]. Field results indicated that even after the operation of open and closed loop power control, the received signal level on the IS-95 system reverse channel varies according to a log-normal distribution with a variance of 1-2 dB. The BER for this model is calculated from the method of [13]. The results of Figure 3 indicate that normal variance of received power levels can result in capacity losses of 10-30% in a CDMA system. Simpson and Holtzman [17] have found that power control can be effectively combined with error correction coding to produce good performance in CDMA systems. At high vehicle speeds, fade durations are sufficiently fast so that the combination of coding and interleaving results in good performance. At lower speeds, the fade durations are longer, allowing the power control scheme time to adjust.

62

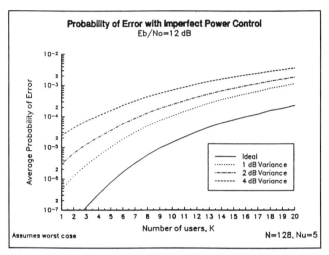

Figure 3: Probability of Error vs. Number of Users for Imperfect Power Control

3.5 Multipath Propagation

Thus far we have assumed that each user's signal is transmitted via a single path. In fact, the wireless channel is characterized by multipath propagation as signals reflect and scatter off of objects in the transmission path. The constructive and destructive interference between arriving multipath components can result in severe fading. The ability of spread-spectrum signals to reject and even exploit multipath propagation is a key advantage in the wireless environment. The typical approach is to model the multipath channel as a time-varying linear filter with impulse response $h(t,\tau)$ given by [18]

$$h_k(t,\tau) = \sum_{l=0}^{L(t)-1} A_l(t)\delta\left(\tau - \tau_l(t)\right)e^{j\theta_l(t)}, \qquad (8)$$

where $L(t)$ is the number of arriving multipath components, $A_l(t)$ is the magnitude of the lth arriving multipath component, $\tau_l(t)$ is the relative multipath delay of the lth arriving component compared with the first arriving component which arrives at $\tau = 0$, and $\theta_l(t)$ is the relative phase of the lth arriving component. The variable τ represents the excess time delay of the multipath in the channel at a specific time t. Each of the channel parameters will vary as a function of time t, depending on the physical movement of the transmitter, receiver, or objects in the channel.

In the presence of multipath propagation, the received signal now may be written as

$$r(t) = \sum_{k=1}^{K} s_k(t - \tau_k) \otimes h_k(t,\tau) + n(t), \qquad (9)$$

where signals are delayed by a uniformly distributed delay τ_k and convolved with the channel impulse response $h_k(t, \tau)$. For the forward channel, all signals emanate from the same antenna and we may assume that the delays $\{\tau_k\}$ and impulse responses $\{h_k(t)\}$ are identical. For the reverse channel the delays and impulse response must be modeled as independent. A tremendous amount of research has focused on the modeling of the channel impulse response, and an excellent overview is contained in [19].

The ability of spread-spectrum to reject multipath signals lies in the autocorrelation properties of the received signals. Multipath components which arrive delayed by more than one chip result in uncorrelated interference rather than destructive intersymbol interference. The shorter the chip duration, the greater the ability to resist multipath fading. In the frequency bands of interest for cellular and PCS, a spreading bandwidth of 1 MHz provides significant resistance to fading and 10 MHz is sufficient to essentially eliminate all fading [9]. Performance of CDMA systems in multipath propagation may be modeled by treating multipath components as additional interferers.

In addition to mitigating the multipath propagation, a spread-spectrum system may actually exploit this effect by using a RAKE receiver [4]. A RAKE receiver is a device for coherently combining the energy from two or more received multipath components, thereby increasing received signal power and providing a form of diversity reception. Although Turin's proposed RAKE receiver [4] was implemented as a tapped delay line with one tap per chip, good performance can usually be obtained with only a few fingers, each synchronized to one multipath component.

3.6 Cellular Capacity for Spread-Spectrum

Although spread-spectrum is inherently well-suited for multipath environments and affords a number of advantages in the areas of networking and hand-off, the key characteristic underlying the current interest in CDMA for wireless cellular systems is the potential for increased spectral efficiency. The argument for capacity of a CDMA system is presented in [3], which employs a Gaussian approximation and a simple path loss model to demonstrate capacity improvements from two key issues. First, the use of CDMA allows improved frequency reuse. Narrowband systems cannot use the same transmission frequency in adjacent cells because of the potential for interference. CDMA has inherent resistance to interference. Although users from adjacent cells will contribute to the total interference level, their contribution will be significantly less than the interference from same cell users. The result is that frequency reuse efficiency increases by a factor of 4-6. In addition, when used to transmit voice signals, CDMA systems may exploit the fact that voice activity typically lies at somewhat less than 40%. As a result, CDMA systems may use statistical multiplexing of voice signals to further increase the capacity.

4. Applications of Spread Spectrum

As stated earlier, spread spectrum found its original home in the military. However, recently it has been proposed for several various civilian applications including cellular telephone, personal communications, and position location [20]. In this section we discuss two applications of CDMA technology which are important and illustrate the properties of spread spectrum.

64

4.1 Cellular Telephone

The Electronics Industry Association has formally adopted a wideband digital standard pioneered by Qualcomm Corporation named IS-95 (Interim standard 95) [2]. This standard employs direct sequence spread spectrum in order to provide dramatically higher capacity when compared to existing analog systems, and possibly more than proposed TDMA systems.

The system uses a 1.2288 Mcps code spreading sequence on top of a variable bit rate that ranges from 1200bps to 9600bps. Two levels of spreading codes are employed. A long code with a period of over one century is used to create a number of potential spreading codes, partly in reaction to the concern that the number of unique codes in more common sets such a m-sequences or Gold codes was too small. A second short code is also used for spreading to enable convenient synchronization. Identical data is transmitted over I and Q channels using different short codes. Simultaneous transmission over the I and Q channel allows extremely sharp pulse shaping.

There are important differences between the forward and reverse links in the IS-95 standard. The performance of the reverse link is of greater concern for two reasons. First, as discussed in Section 3, the reverse link is subject to near/far effects. Second, since all transmissions on the forward link originate at the same base station, it is possible to generate synchronous orthogonal signals (although multipath channels will still cause some multiple access interference). For this reason, more powerful error correction is employed on the reverse link. Data is encoded using a rate 1/2 constraint length 9 convolutional code followed by an interleaver on the forward channel, while a rate 1/3 constraint length 9 convolutional code followed by an interleaver is used on the reverse link. Furthermore, on the reverse link a Walsh code is used to modulate the data, providing another level of error correction. Interleaving is utilized to avoid large burst errors which can be very detrimental to convolutional codes. The signal generation process for the forward and reverse channels is shown in more detail in Figures 4 and 5 respectively [21]. (QPSK modulation is implied but not explicitly shown.)

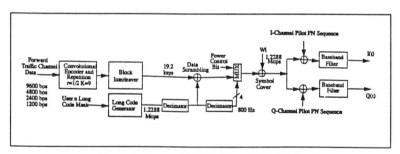

Figure 4: Forward CDMA Channel Modulation Process

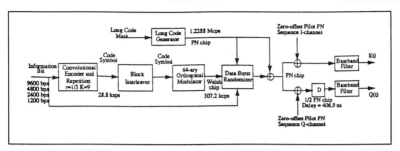

Figure 5: Reverse CDMA Channel Modulation Process

As mentioned earlier, the near-far problem needs to be addressed when spread spectrum is utilized in mobile communications. To combat this problem IS-95 uses power control. On the forward link there is a subchannel for power control purposes. Every 1.25ms the base station receiver estimates the signal strength of the mobile unit. If it is too high, the base transmits a '1' on the subchannel. If it is too low, it transmits a '0'. In this way the mobile adjusts its power every 1.25ms as necessary so as to reduce interference to other users.

The QCELP Linear Predictive Coding algorithm is used for voice encoding. Since the voice coder exploits gaps and pauses in speech, the data rate is variable. In order to keep the symbol rate constant, whenever the bit rate falls below the peak bit rate of 9600 kbps, repetition is used to fill the gaps. For example, if the output of the voice coder (and subsequently the convolutional coder) falls to 2400bps, the output is repeated three times before it is sent to the interleaver. IS-95 takes advantage of this repetition time by reducing the output power during three out of the four identical symbols by at least 20dB. In this way, the multiple access interference is reduced. This voice activity gating reduces interference and increases overall capacity.

The performance of the proposed system is simulated in [21]. The simulated frame error rate vs. the number of users per channel (1.25 MHz) for a typical suburban environment is shown in Figure 6. The simulation of CDMA systems is inherently computationally intensive because the required sampling rate is directly proportional to the bandwidth. As a result, semianalytic techniques such as importance sampling are becoming increasingly necessary for performance evaluation [22,23]. Figure 6 also shows the same simulations repeated using importance sampling techniques from [24], requiring an order of magnitude less computer time.

66

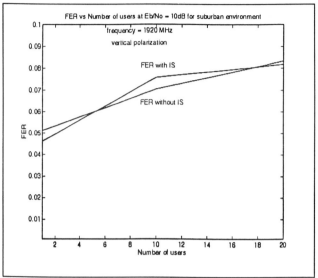

**Figure 6: FER vs. Number of Users at Eb/No = 10dB for forward channel
in a suburban environment at 1920 MHz with vertical polarization**

Another application of spread spectrum in cellular telephony is Broadband CDMA [25,9]. Broadband CDMA is a technique introduced by InterDigital Communications Corp. which allows overlay of the current AMPS with a CDMA system. In this way, the system can be smoothly converted from analog to digital without displacing any users. The interference between the two systems is inherently low due to the noise-like properties of CDMA. However, to insure sufficiently low C/I ratios, notch filters are placed in the CDMA transmitters and receivers. These 30kHz notch filters are placed at several frequencies where AMPS users are located. The Broadband CDMA approach features a less complex signaling scheme than the IS-95 standard. However, the broadband CDMA signal is inherently more resistant to fading as discussed in Section 3.

4.2 Position Location

One current application of spread spectrum techniques is the Global Positioning System or GPS. GPS as its name implies, is a satellite-based position location system and is illustrated in Figure 7. GPS like spread spectrum found its original application in the military. GPS however, is now finding several uses in civilian life such as ship and aircraft navigation, surveying, and geological studies [26].

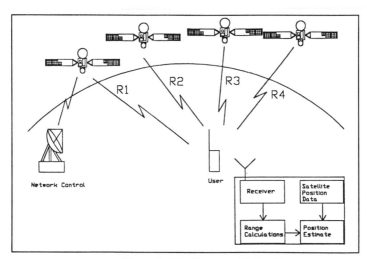

Figure 7: The Global Positioning System

GPS uses 24 satellites at a height of 22,200 km located in six orbital planes. Each satellite circles the earth in 12 hours and emits two pseudorandom noise (PN) sequences modulated in phase quadrature at two frequencies. Two frequencies are needed to correct for the delay induced by the ionosphere. The range to a satellite is calculated by receiving a short pulse and comparing the time difference between a reference clock and the received pulse. A user locates their position by receiving the signal from four of the possible 24 satellites. There are four unknowns, the co-ordinates in three-dimensional space of the user along with a timing bias within the user's receiver. These four unknowns can be solved by using four range equations to each of the four satellites. The accuracy of this system is related to the pulse duration. Thus, the smaller the pulse the better the resolution.

The two PN sequences are called the P code (precision code) and the C/A code (coarse acquisition code). The C/A code is 1023 chips long repeating every millisecond (the chip rate is approximately 1 Mbps) and can be locked onto via a sliding correlator fairly easily (within about 0.5 seconds). The P code however is significantly longer with each satellite's P code repeating about every 7 days and has a chip rate of 10Mbps. This code is much more difficult to lock onto and is only available to select users. This is due to the theoretical improved accuracy provided by the P code's shorter pulse duration. The theoretical accuracy of the P code is 0.6m while the theoretical accuracy of the C/A code is 6m. Realistic accuracies are 6m and 60m respectively. (Ironically, measurements have shown that the C/A code can actually perform nearly as well as the P code due to the high performance of commercially available receivers).

The use of spread spectrum in this system accomplishes three things. First, as mentioned earlier, the signals from the satellites (specifically the P code) can be kept from unauthorized use. Second and more importantly from a practical sense, is that the inherent processing gain of spread spectrum allows reasonable power levels to be used. Since the cost of a satellite is proportional to its weight, we need to reduce the power required as much as possible. Additionally, since each satellite must see the entire hemisphere, very little antenna gain is possible. For high accuracy short pulses are required to provide fine resolution. This results in high spectrum occupancy. The result is that the received signal is several dB below the noise floor. Since range information needs to be calculated only about once per second, the data bandwidth need only be around 100 Hz. This is a natural match for spread spectrum. By compressing the received signal in the receiver, a significant processing gain is realized, thus allowing good reception at reasonable power levels. The third reason for SS is that each satellite can use the same frequency band without interfering with each other due to the near orthogonality of each user's signal.

Spread spectrum is also being used for terrestrial based position location in Automatic Vehicle Monitoring (AVM) systems [27,28].

5. Research Issues in Spread Spectrum

A wide range of active research areas remain in the field of spread spectrum communications. Rapid acquisition and synchronization techniques will become increasingly important as CDMA data networks are deployed [10]. The application of directional and adaptive antenna technology to CDMA systems has roused considerable interest [29,30]. Gaussian interference models have shown that even modest receiver directionality at the base station can result in substantial capacity improvements [31]. In this section we focus on two active areas of research: receiver design for interference rejection and the choice of error correction coding techniques for CDMA systems.

5.1 Receiver Design for Interference Rejection.

The current generation of CDMA systems employ single stage correlation receivers which correlate the received signal with a synchronized copy of the desired signal's spreading code. One improvement on this technology is the RAKE receiver which combines information from multiple resolvable signal components [4]. For the current receivers, the capacity claims of CDMA are based entirely on frequency reuse in a multi-cell system. If only a single cell is considered, CDMA systems employing simple correlation receivers cannot approach the spectral efficiency of time or frequency division multiple access. Furthermore, correlation and RAKE receivers are particularly susceptible to the near/far problem. As discussed in Section 3, even if sophisticated power control is employed, near/far effects can result in significant performance degradation.

By viewing the receiver design in terms of multi-user rather than single-user reception, the capacity of a CDMA system can be increased by an order of magnitude or more. This approach is particularly applicable at the base station, which must simultaneously receive signals from all users, and which has significant processing power available. Optimal multi-user receivers for CDMA systems employ a variation of the Viterbi algorithm to simultaneously demodulate all received signals [32]. For realistic numbers of users, such optimal receivers are impractical; however, a sub-optimal approach can achieve impressive results.

Multistage correlation receivers, due to Varanasi and Aazhang [33], are one promising approach. At each stage, a bank of single user receivers demodulates the received signal. After each stage, the estimated signals for each interferer are subtracted from the received signal, and then demodulation is repeated. This can be repeated for an arbitrary number of stages to obtain an iterative estimate of the interference, but results for simple channel models indicate that only a few stages are necessary to achieve most of the potential performance improvement. This receiver configuration is also highly resistant to the near/far problem; however, the work of [33] assumes idealized channel models and perfect knowledge of each user's power, so it is not directly applicable to wireless systems.

One approach is to adapt the multistage receiver so that it simultaneously forms estimates of channel and the interference characteristics. Use of a RAKE receiver in a multistage algorithm makes this possible. The RAKE receiver demodulates the strongest received components in a CDMA systems which may be used to form an estimate of the channel impulse response for each user. A possible multistage RAKE receiver, illustrated in Figure 8, uses the output of a RAKE receiver to simultaneously estimate the received data and the channel impulse response for each user. The estimated signal is then reconstructed and subtracted from the received signal. The new received signal is used to perform another stage of estimation. This approach allows the multistage receiver to form iterative estimates of both the channel and the received data, and allows the multistage receiver to operate with no prior information about channel conditions. Simulated performance of a multistage RAKE in the presence of multipath propagation and imperfect power control are presented in Figure 9 [34]. A two stage RAKE receiver with seven users has performance roughly equivalent to a conventional RAKE receiver with two users. Several other researchers are currently pursuing similar approaches [35,36]. The iterative receiver design is just one of several interference rejection techniques available. A thorough survey of these techniques is presented in [37].

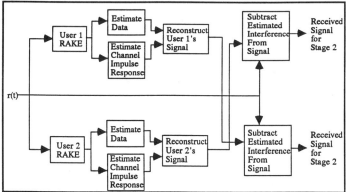

Figure 8: Block Diagram of a Multistage RAKE Receiver

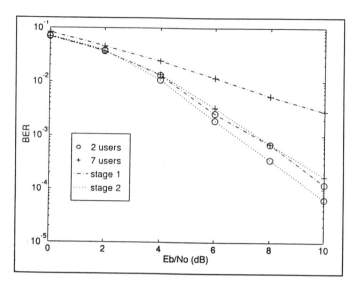

Figure 9: BER as a function of Eb/No for a 2 stage RAKE for K=2 and 7 in the presence of 2 multipath components and log-normal signal power distribution with variance 1 dB

5.2 Error Correction for Spread-Spectrum

Because of the multiple access interference, all CDMA systems require forward error correction for reliable performance. There has been a great deal of theoretical work on error correction coding for spread spectrum systems in the past decade. Today's commercial CDMA systems favor standard convolutional codes. For example, the I-95 CDMA system employs a rate 1/2, constraint length 9 code for the forward channel and a rate 1/3, constraint length 9 code for the reverse channel. Two other coding techniques hold promise for use in future systems.

Trellis codes, first proposed by Gottfried Ungerboeck [38], combine coding and modulation into a single operation. Trellis coding techniques have been applied successfully to narrowband channels such as telephone lines, satellite and microwave lines to dramatically increase the data rate of modems for bandwidth constrained systems. Trellis coding offers one potential way of improving the performance of CDMA wireless systems. Early research discovered that trellis codes based on PSK signal sets do not outperform standard convolutional codes when applied to a CDMA system [39]. Recent research, however, indicates that when the trellis codes are based on biorthogonal signal sets, the performance of a CDMA system can be significantly improved [40]. That is, the trellis code for a CDMA system combines the error correction coding and choice of spreading code into a single operation. Thus, while a standard

CDMA system transmits one of two spreading sequences, a trellis coded CDMA system transmits one of many possible spreading sequences.

Another option for improved error correction in a CDMA system is the use of very low rate convolutional codes. By very low rate, we mean 1/16 or 1/64, rather than typical rates like 1/2. There is a relatively simple way of generating these very low rate codes known as orthogonal convolutional codes [41]. Viterbi has conjectured that very low rate convolutional codes will give optimum performance for a CDMA system [42]. Figure 10, shows a comparison of low rate convolutional codes with trellis codes for CDMA systems [43]. The two codes offer comparable performance, with lightly loaded situations slightly favoring the trellis codes and heavily loaded situations favoring the convolutional code.

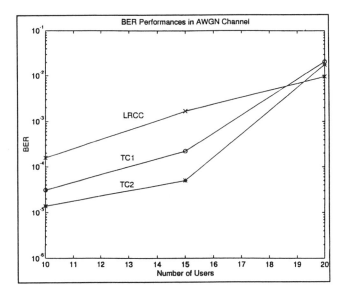

**Figure 10: BER Performance in AWGN of
Convolutional Coding Combined with Trellis Coding**

6. Conclusions

This paper has presented fundamental principles underlying spread spectrum communications. These principles have been illustrated in terms of common system models and analysis techniques. Both position location and cellular applications of CDMA have been discussed. Finally, several areas of current research in spread spectrum were highlighted. Spread spectrum is a diverse and fascinating field and the interested reader is referred to the wide range of referenced work for more complete details.

Acknowledgment

This work was supported in part by the Bradley Fellowship program.

References

[1]   R.A. Scholtz, "The Origins of Spread Spectrum Communications," *IEEE Transactions on Communications*, Vol. COM-30, N0. 5, May 1982.

[2]   Committee on Digital CDMA Cellular, *IS-95 Wideband Spread Digital Cellular System Dual Mode Mobile Station-Base Station Compatibility Standard*, Technical Report, EIA?TIA, TR45.5, April 1992.

[3]   K.S. Gilhousen, I.M. Jacobs, A.J. Viterbi, et al, "On the Capacity of a Cellular CDMA System," *IEEE Transactiopns on Vehicular Technology*, Vol. 40, No. 2, May 1991.

[4]   G. L. Turin, "Introduction to Spread-Spectrum Antimultipath Techniques and Their Application To Urban Digital Radio," *Proceedings of the IEEE*, Vol. 68, No. 3, March 1980.

[5]   M.B. Pursley, "Performance Analysis for Phase-Coded Spread-Spectrum Multiple-Access Communication-Part I: System Analysis," *IEEE Transactions on Communications*, Vol. COM-25, No. 8, August 1977.

[6]   G.R. Cooper, R.W. Nettleton, and D.P. Grybos, "Cellular Land-Mobile Radio: Why Spread Spectrum?," *IEEE Communiations Magazine*, March 1979.

[7]   R.C. Dixon, Spread Spectrum Systems, John Wiley & Sons, 1984.

[8]   R.E. Ziemer and R.L. Peterson, Digital Communication and Spread Spectrum Systems, Macmillan, 1985.

[9]   R.L. Pickholtz, L.B. Milstein and D.L. Schilling, "Spread Spectrum for Mobile Communications," *IEEE Transactions on Vehicular Technology*, Vol. 40, No. 2, May 1991.

[10] The AARL Spread Spectrum Sourcebook, Edited by A. Kestleloot and C.L. Hutchinson, The American Radio Relay League.

[11] D.V. Sarwate and M.B. Pursley, "Crosscorrelation Properties of psuedorandom and related sequences," *Proceedings of the IEEE*, Vol. 68, No. 5, May 1980.

[12] M.B. Pursley and E. Geraniotis, "Error Probabilities for Direct Sequence Spread Spectrum Multiple Access Communications - Part II: Approximations," *IEEE Transactions on Communications*, May 1982.

[13] J.S. Lehnert, "An Efficient Technique for Evaluating Direct-Sequence Spread-Spectrum Communications," IEEE Transactions on Communications, Vol. 37, No. 8, August 1989.

[14] R.K. Morrow and J.S. Lehnert, "Bit-to-Bit Error Dependence in Slotted DS/SSMA Packet Systems with Random Signal Sequences," *IEEE Trans. on Comm*, vol. COM-37, pp.1052-1061, Oct. 1989.

[15] J. M. Holtzman, "A Simple Accurate Method to Calculate Spread-Spectrum Multiple Access Error Probabilities," *IEEE Trans. on Comm.*, vol COM-40, pp.461-464, March 1992.

[16] B.D. Woerner and R. Cameron, "An Analysis of CDMA with Imperfect Power Control," IEEE Vehicular Technology Conference, May 1992.

[17] F. Simpson and J. Holtzman, "CDMA Power Control, Interleaving and Coding," *IEEE Veh. Tech. Conf.* (St. Louis, MO), pp. 362-367, May 1991.

[18] G. Turin, "A Statistical Model of Urban Multipath Propagation", *IEEE Trans. on Veh. Technology*, vol. VT-21, no. 1, pp. 1-11, Feb. 1972.

[19]   H. Hashemi, "Impulse Response Modeling of Indoor Radio Propagation Channels," *IEEE Journal on Selected Areas in Communications,* Vol. 11, No. 7, Sept. 1993

[20]   D.L. Schilling, R.L. Pickholtz, and L.B. Milstein, "Spread Spectrum Goes Commercial," *IEEE Spectrum,* August 1989.

[21]   Y.L. Li, B.D. Woerner, W. Tanis II, and M. Hughes, "Simulation of CDMA Using Measured Channle Impulse Response Data," *IEEE Veh. Tech. Conf.* (Secaucus, NJ), May 1993.

[22]   M. Jeruchim, P. Balaban, and K.S. Shannmugan, *Simulation of Communication Systems,* Plenum Press, New York, 1992.

[23]   W.A. Al-Qaq, M. Devetsikiotis and J.K. Townsend, "Importance Sampling Methodologies for Simulation of Wireless Communication Links," Third Annual Virginia Tech Symposium on Wireless Personal Communications, June 1993.

[24]   S. Nagpal, "Application of Importance Sampling Simulation to CDMA Systems," *M.S. Thesis,* Virginia Polytechnic Institute and State University, Blacksburg, VA, 1984.

[25]   G.R. Lomp, D.L. Schilling, and L.B. Milstein, "An Overview of Broadband CDMA," Virginia Tech's Fourth Symposium on Wireless Personal Communications, June 1-3, 1994.

[26]   I.A. Getting, "The Global Positioning System," *IEEE Spectrum,* Vol. 30, No. 12, December 1993.

[27]   R. L. Pickholtz, "Engineering Analysis of Cochannel Pulse-Rangin LMS Systems," Appendix to Comments of North American Teletrac and Location Technologies, Inc., PR Docket No. 93-61 RM-8013, June 28, 1993.

[28]   D. J. Torrieri, "Statistical Theory of Passive Location Systems," *IEEE Trans. on Aerospace and Electronic Systems,* Vol. AES-12, No. 2, March 1984.

[29]   S. C. Swales, et. al., "The Performance Enhancement of Multibeam Adaptive Base-Station Antennas for Cellular Land Mobile Radio Systems," *IEEE Trans. Veh. Tech.,* vol. VT-39, no. 1, pp. 56-67, Feb. 1990.

[30]   B. G. Agee, K. Cohen, J. H. Reed, T. C. Hsia, "Simulation Performance of A Blind Adaptive Array For a Realistic Mobile Channel", *43rd IEEE Vehicular Technology Conference,* Meadowlands, NJ, May, 1993, pp. 97-100.

[31]   J. C. Liberti and T.S. Rappaport, "Reverse Channel Performance Improvement in CDMA Cellular Communication Systems Employing Adaptive Antennas," *Proceedings IEEE Globecom '93,* Houston, pp.42-47, Dec. 1993.

[32]   S. Verdu, "Minimum Probability of Error for Asynchronous Gaussian MultipleAccess Channel," *IEEE Transactions on Information Theory,* Vol. IT-32, No. 5, Sept. 1986.

[33]   M.K. Varanasi and B. Aazhang, "Multisatge Detection in Asynchronous Code Division Multiple Access Communications," *IEEE Transactions on Communications,* Vol. COM-38, No. 4, April 1990.

[34]   S. Striglis, A. Kaul, N. Yang, and B. Woerner, "A Multistage RAKE Receiver for Improved CDMA Performance," IEEE Vehicular Technology Conference, May 1994.

[35]   S. S. H. Wijayasuriya, J. P. McGeehan and G.H. Norton, "RAKE Decorrelation as an Alternatlve to Rapid Power Control in DS-CDMA Mobile Radio," *IEEE Vehicular Technology Conference*   (VTC), Piscataway, NJ, May 1993.

[36]   P. Patel and J.M. Holtzman, "Analysis fo DS/CDMA Successive Interference Cancellation Scheme Using Correlation," *Globecom '93,* December 1993.

[37]   R.D. Holley and J. H. Reed, "Time dependent adaptive filters for interference cancelation in CDMA systems," unpublished master's thesis, Mobile and Portable Radio Research Group (MPRG-TR-93-15A), Bradley Department of Electrical Engineering, Virginia Polytechnic and State University, October 1993.

[38]   G. Ungerboeck, "Channel coding with multilevel/phase signals," *IEEE Transactions on Information Theory,* vol. IT-40, no. 1, pp. 55-67, January 1982.

74

[39]   G. D. Boudreau, D. D. Falconer and S. A. Mahmoud, "A comparison of trellis coded versus convolutionally coded spread-spectrum multiple access systems," *IEEE Journal on Selected Areas in Communications*, Vol. 8, no. 4, pp. 628-640, May 1990.

[40]   B. D. Woerner and Wayne E. Stark, "Trellis coding for direct-sequence spread-spectrum communications," *IEEE Transactions on Communications*, accepted for publication.

[41]   A. J. Viterbi and J. K. Omura, Principles of Digital Communication and Coding, New York: McGraw-Hill, 1979.

[42]   A. J. Viterbi, "Very low rate convolutional codes for maximum theoretical performance of spread-spectrum multiple-access channels," *IEEE Journal on Selected Areas in Communications*, Vol. 8, no. 4, pp. 641-649, May 1990.

[43]   Y.M. Kim and B. D. Woerner, "Comparison of Low Rate Trellis Coding versus Low Rate Convolutional Codes for CDMA Systems," *submitted to IEEE MILCOM '94.*

4

CMA Adaptive Array Antenna Using Transversal Filters for Spatial and Temporal Adaptability in Mobile Communications

Nobuyoshi KIKUMA   Kazuya HACHITORI   Fuminobu SAITO
Naoki INAGAKI

Department of Electrical and Computer Engineering
Nagoya Institute of Technology
Nagoya 466, Japan

Abstract

The CMA (Constant Modulus Algorithm) was developed for the adaptive receiving systems to capture the desired signal having the constant modulus property such as FM, PSK and FSK signals. The CMA is an algorithm suitable for mobile communications because it has the great advantage of not requiring a reference signal as the LMS algorithm does.

The CMA adaptive transversal filter can equalize channel distortions affecting the data signal received in multipath environments. However, it cannot eliminate the influence of the co-channel interferences that are not correlated with the desired signal. On the other hand, the CMA adaptive array antenna can suppress the co-channel interferences as well as the multipath signals. Therefore, this paper shows the performance of the CMA adaptive array antenna using the transversal filters (tapped delay lines) in the presence of the multipath signals and co-channel interference. It can be regarded as the spatial and temporal adaptive signal processor.

In this paper, computer simulation is carried out using an antenna array equipped with tapped delay lines for the antenna weights. A $\pi/4$-shifted QPSK signal is generated which is transmitted over several multipath channels. Furthermore, another signal of the same modulation type is generated for the co-channel interference.

For optimization of the CMA adaptive system, the steepest descent method and Marquardt method are utilized. The former has mainly been used for the CMA because the cost function is nonlinear with respect to the tap weights. However, the slow convergence of the algorithm has often limited its application in mobile communications where signals must be quickly captured. The latter, on the other hand, is one of the nonlinear least squares algorithms to attain its rapidly-converging and well-conditioned adaptation.

The simulation results demonstrated that the spatial and temporal adaptive system can achieve high quality of communications with low BERs. Also, it is shown that the Marquardt method can contribute to reducing the convergence time.

1. Introduction

The efficient utilization of frequency spectrum is still one of the important topics in the field of radio communications. From this viewpoint, the receiving systems are required to reject the unwanted waves in a complicated radio environment. Adaptive array antennas are fascinating system that can suppress the unwanted signals while maintaining the desired signal at the array output. We have strongly been proposing the application of the adaptive array antennas to mobile communications which are attracting public notice nowadays.

The CMA (Constant Modulus Algorithm) was developed for the adaptive receiving systems to capture the desired signal having the constant envelope property such as FM, PSK and FSK signals[1]-[7]. The CMA is an adaptive algorithm suitable for mobile communications which are characterized by multipath environments. This is because it has the great advantage of not requiring a priori knowledge about the transmitted signal.

The CMA adaptive filter with a tapped delay line can equalize channel distortions affecting the data signal received in multipath environments[1]. However, it cannot eliminate the influence of the co-channel interferences that are not correlated with the desired signal. On the other hand, the CMA adaptive array antenna can suppress the co-channel interferences as well as the multipath signals[2]-[6]. In this paper, therefore, we show via computer simulation the performance of the CMA adaptive array antenna using the tapped delay lines in the presence of the multipath signals and co-channel interference. It can be regarded as the spatial and temporal adaptive signal processor[9][10].

2. CMA Adaptive Array with Tapped Delay Lines

2.1 Configuration and Principle

We consider a K-element antenna array whose channels are equipped with L-tapped delay lines for the antenna weighting. Figure 1 shows a configuration of the array receiving system. Let x_i and w_i represent the input and weight at the ith tap point ($i = 1, \cdots, KL$) respectively, and also X and W denote the input vector and weight vector, respectively, which are defined as

$$X = [x_1, \cdots, x_{KL}]^T \qquad (1)$$
$$W = [w_1, \cdots, w_{KL}]^T \qquad (2)$$

Then, the array output y is expressed as

$$y = X^T W^* \qquad (3)$$

The CMA adaptive array works to eliminate the amplitude fluctuations of the array output signal due to the incidence of interferences. Therefore, the cost function to be minimized is normally represented as

$$Q(W) = E\left[\left(|y|^2 - \sigma^2\right)^2\right] \qquad (4)$$

where $E[\cdot]$ denotes the ensemble mean and σ is the amplitude of the array output signal expected in the absence of signal degradation[1].

2.2 Optimization Algorithm

In this paper, we adopt two algorithms for the CMA optimization, which are the steepest descent method and Marquardt method[4]-[6][8]. They are described below.

Steepest descent method

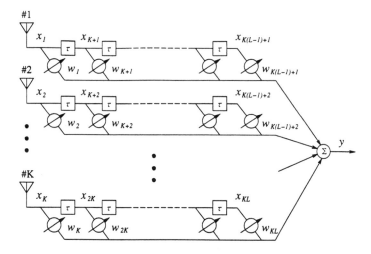

x : input, y : output, w : weight, τ : time delay

Figure 1: Configuration of of a adaptive array with tapped delay lines

The steepest descent method is often adopted to iteratively and asymptotically converge the weight vector to its optimum. For optimization of the CMA adaptive array, the steepest descent method has mainly been used because the cost function is nonlinear with respect to the weights. The basic algorithm for the CMA can be expressed as follows:

$$
\begin{aligned}
\boldsymbol{W}_{L+1} &= \boldsymbol{W}_L - \mu \nabla_W Q(\boldsymbol{W}_L) \\
&= \boldsymbol{W}_L - 4\mu \boldsymbol{X}_L y_L^*(|y_L|^2 - \sigma^2)
\end{aligned}
\tag{5}
$$

where L is an iteration number for the weight update and μ is a parameter called a step size. It is noted that the ensemble mean is removed in eq.(5). In the practical operation, the average obtained from the finite sample data is used for the ensemble mean, as is often the case.

Marquardt method

The slow convergence of the steepest descent method has often limited its application in mobile communications where signals must be quickly captured[5]. The Marquardt method, on the other hand, is one of the nonlinear least squares techniques to attain its rapidly-converging and well-conditioned adaptation. Thus, we explain here the optimization algorithm based on the Marquardt method.

To obtain the optimum weight vector, we minimize $\sum_{i=1}^{m}(|\boldsymbol{X}_i^T \boldsymbol{W}^*|^2 - \sigma^2)^2$ for a set of m input data ($\boldsymbol{X}_1, \boldsymbol{X}_2,..., \boldsymbol{X}_m$). On the minimization, the Marquardt method updates $\boldsymbol{W}$ by the offset $\Delta \boldsymbol{W}$ in the following manner:

$$
\boldsymbol{W}_{L+1} = \boldsymbol{W}_L + \Delta \boldsymbol{W}_L
\tag{6}
$$

$$(J_L^T J_L^* + \alpha U)\Delta W_L = -J_L^T H(W_L) \tag{7}$$

$$H(W_L) = [h_1(W_L), h_2(W_L), ..., h_m(W_L)]^T \tag{8}$$

$$J_L = [\nabla h_1(W_L), \nabla h_2(W_L), ..., \nabla h_m(W_L)]^T \tag{9}$$

$$h_i(W_L) = |X_i^T W_L^*|^2 - \sigma^2 \qquad (i = 1, 2, ..., m) \tag{10}$$

$$\nabla h_i(W_L) = X_i X_i^\dagger W_L \qquad (i = 1, 2, ..., m) \tag{11}$$

where U denotes the identity matrix, and α is a positive parameter called Marquardt number which permits a satisfactory convergence. The particular case of $\alpha = 0$ is equivalent to the Newton-Gauss method[7]. On the other hand, the case where α is large enough approximates to the steepest descent method. Therefore, the Marquardt method is regarded as the mixture of the Newton-Gauss method and the steepest descent method. Although the optimum value of α must be searched out at each iteration, we definitely choose for α the squared norm of Jacobian matrix J_L as follows:

$$\alpha = \frac{\sum_{i=1}^m ||\nabla h_i(W_L)||^2}{(mK)^2} \tag{12}$$

Thus, α is very large when W is far from the optimum point. As W approaches the optimum point, α decreases and accordingly the Marquardt algorithm behaves like the Newton-Gauss algorithm near the optimum point[5].

3. Computer Simulation

We show the results of computer simulation using a linear array of isotropic antenna elements. The element spacing of the array is a half wavelength of the carrier frequency and the tap spacing is a symbol duration $(= T)$ of the transmitted data signal. We generate a $\pi/4$-shifted QPSK signal which is transmitted over several multipath channels. In addition, another signal of the same modulation is generated for the co-channel interference. Sampling rate is $8/T$, and all the signals are not filtered and not disturbed by channel fading. The input SNR (Signal-to-Noise Ratio) is defined as the ratio of the power of the first arriving wave to the power of the thermal noise at each antenna element. In the CMA optimization, we let $\sigma = 1$ and use 15 snapshots $(m = 15)$ for one update of weight vector both in the steepest descent method and the Marquardt method on condition that each snapshot is taken at the center of symbol duration of the first arriving signal. The initial weight vector is set to $W_0 = [1, 0, \cdots, 0]^T$, which means the initial array pattern is isotropic.

3.1 2-Wave Model (Scenario 1)

Table 1 gives the detail of radio environment used in simulation experiment for a 2-wave multipath model. Angles of arrival are measured from the broadside direction of array.

Steepest descent method

For optimization of the CMA adaptive system, we first utilized the steepest descent method.

Table 1: Radio environment used in computer simulation (Scenario 1)

	power	angle of arrival	delay time
wave 1	0 dB	0°	0
wave 2	–2 dB	60°	T
Input SNR = 20 dB			

(T : symbol duration of $\pi/4$-shifted QPSK signal)

Figure 2 shows the relation of average BER (Bit Error Rate) to the iteration of weight update in the case of 1-element system which is equivalent to the conventional adaptive equalizer. Four lines in this figure demonstrate that the system with the more taps brings out the better BER performance. Furthermore, the behavior of 2-element system is shown in Fig. 3. It is found that the 2-element system gives better performance than the 1-element system. It is because the 2-element system suppresses the second wave by creating a null of the array pattern in the direction of the second wave. To examine the characteristics from the viewpoint of an array system, we present the array output powers versus the iteration and the directional patterns after convergence. Figures 4 and 5 show the characteristics of the 2-element, 1-tap system and the 2-element, 2-tap system, respectively. In Fig. 4, it is observed that the system makes a null in the direction of the second wave. In Fig. 5, it is noted that the second wave in the first tap array (virtual array composed only of the first tap signals) is canceled by the first wave in the second tap array. Thus, only the first wave in the first tap array is preserved at the array output as a result of the spatial and temporal signal processing. The 3-tap and 4-tap systems of 2-element array have the same characteristics as the 2-tap system, which are not shown here.

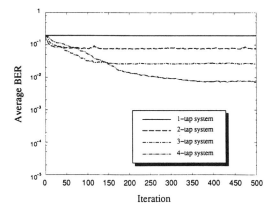

Figure 2: The relation of average BER to iteration of the 1-element system with the steepest descent method (Table 1)

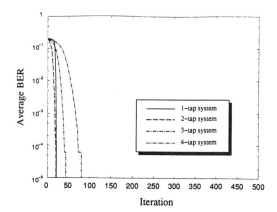

Figure 3: The relation of average BER to iteration of the 2-element system with the steepest descent method (Table 1)

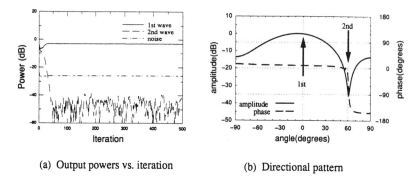

(a) Output powers vs. iteration

(b) Directional pattern

Figure 4: Array output powers versus the iteration and the directional patterns after convergence of the 2-element, 1-tap system with the steepest descent method (Table 1)

Marquardt method

Next, we utilized the Marquardt method for optimization of the CMA adaptive system. The behavior of the adaptive system with the Marquardt method is basically similar to the steepest descent method. The difference between them is that the Marquardt method gives much faster rate of convergence. To demonstrate that, we show the average BER versus the iteration of the 2-element system with the Marquardt method in Fig. 6. It is found from the figure that the Marquardt method attains much higher rate of convergence than the steepest descent method shown in Fig.3.

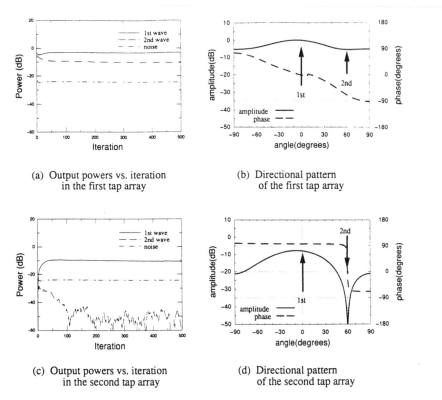

(a) Output powers vs. iteration
in the first tap array

(b) Directional pattern
of the first tap array

(c) Output powers vs. iteration
in the second tap array

(d) Directional pattern
of the second tap array

Figure 5: Array output powers versus the iteration and the directional patterns after convergence of the 2-element, 2-tap system with the steepest descent method (Table 1)

3.2   3-Wave Model (Scenario 2)

Table 2 gives the detail of radio environment used in simulation experiment for a 3-wave model. The first and second waves are multipath propagation waves, and the third wave is a co-channel interference that is independent of the first and second waves.

Table 2: Radio environment used in computer simulation (Scenario 2)

	power	angle of arrival	delay time
wave 1	0 dB	0°	0
wave 2	−2 dB	60°	T
wave 3	−3 dB	−50°	(incoherent)
Input SNR = 20 dB			

(T : symbol duration of $\pi/4$-shifted QPSK signal)

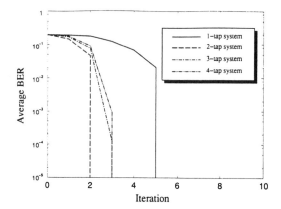

Figure 6: The relation of average BER to iteration of the 2-element system with the Marquardt method (Table 1)

Figure 7 shows the relation of average BER to the iteration of weight update in the case of the 2-element, 2-tap system. In the figure, (a) corresponds to the case of Marquardt method and (b) the case of the steepest descent method. Both methods attain good BER performance, and in particular the Marquardt method realizes much higher rate of convergence than the steepest descent method. The array output powers versus the iteration and the directional patterns after convergence in the case of the Marquardt method are shown in Fig. 8. From this figure, it is found that the third incoherent wave is suppressed both in the first tap and the second tap and also that the second wave in the first tap array is canceled by the first wave in the second tap array as in Fig. 5. In other words, the co-channel interference is suppressed by the spatial processing and the second multipath wave is canceled by the temporal processing. Consequently, only the first wave is maintained at the array output.

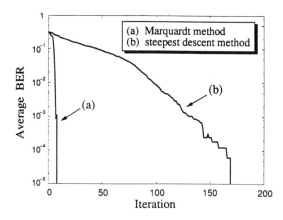

Figure 7: The relation of average BER to iteration of the 2-element, 2-tap system (Table 2)

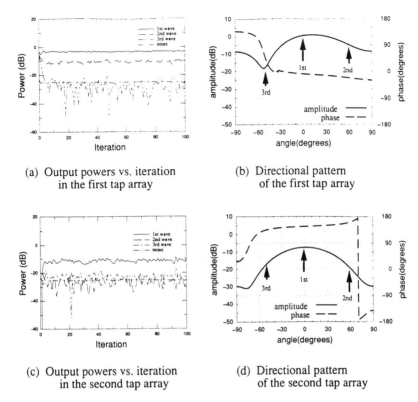

(a) Output powers vs. iteration
in the first tap array

(b) Directional pattern
of the first tap array

(c) Output powers vs. iteration
in the second tap array

(d) Directional pattern
of the second tap array

Figure 8: Array output powers versus the iteration and the directional patterns after convergence of the 2-element, 2-tap system with the Marquardt method (Table 2)

4. Conclusion

In this paper, we investigated the performance of the CMA adaptive array antenna using tapped delay lines in the presence of the multipath signals and co-channel interference. From the configuration, the presented adaptive system can be regarded as the spatial and temporal signal processor. In computer simulation, we employed the steepest descent method and the Marquardt method for the optimization algorithm.

The simulation results have demonstrated that the spatial and temporal CMA adaptive system can achieve high quality of communications against the multipath delayed signal and co-channel interference on both optimization algorithms. Particularly, it is shown that the Marquardt method realizes much higher rate of convergence than the steepest descent method and contributes to quick capture of the desired signal in mobile communications.

References

[1] J.R.Treichler and B.G.Agee, "A New Approach to Multipath Correction of Constant Modulus Signals," IEEE Trans., vol.ASSP-31, pp.459–472, April 1983.

[2] R.Gooch and,J.Lundell, "The CM Array : An Adaptive Beamformer for Constant Modulus Signals," Proc. IEEE ICASSP. pp.2523–2526, 1986.

[3] T.Ohgane, "Characteristics of CMA Adaptive Array for Selective Fading Compensation in Digital Land Mobile Radio Communications," Trans. IEICE of Japan, vol.J73-B-II, pp.489–497, Oct. 1990.

[4] M.Fujimoto, N.Kikuma and N.Inagaki, "Performance of CMA Adaptive Array Optimized by Marquardt Method for Suppressing Multipath Waves," Trans. IEICE of Japan, vol.J74-B-II, pp.599–607, Nov. 1991.

[5] N.Kikuma, M.Fujimoto and N.Inagaki, "Rapid and Stable Optimization of CMA Adaptive Array by Marquardt Method," Proc. IEEE APS International Symposium. vol.1, pp.102–105, June 1991.

[6] M.Fujimoto, "Improvement of Suppression Performance of CMA Adaptive Array to Coherent Interference," Trans. IEICE of Japan, vol.J75-B-II, pp.789–796, Nov. 1992.

[7] B.G.Agee, "The Least-Squares CMA : A New Technique for Rapid Correction of Constant Modulus Signals," Proc. IEEE ICASSP, pp.953–956, 1986.

[8] D.W.Marquardt, "An Algorithm for Least Squares Estimation of Nonlinear Parameters," SIAM J. Appl. Math, No.11, pp.431–441, 1963.

[9] N.Kuroiwa, R.Kohno and H.Imai, "Design of a Directional Diversity Receiver Using an Adaptive Array Antenna," Trans. IEICE of Japan, vol.J73-B-II, pp.755–763, Nov. 1990.

[10] I.Tsujimoto, "Adaptive Equalization and Interference Cancellation by using an Arrayed Decision Feedback Equalizer," Proc. of 1993 IEICE Spring Conf. of Japan, B-339, 1993.

5

The effect of adjacent cell interference, power control and error correction on a CDMA cellular system

A Sathyendran[*], K W Sowerby[*], and M Shafi[#]

[*] Department of Electrical and Electronic Engineering, University of Auckland, New Zealand.

[#] Telecom Corporation of New Zealand Ltd, New Zealand.

Email : a.sathyendran@auckland.ac.nz

Abstract

The performance of a CDMA cellular radio system can be improved by employing power control and error correction. This paper discusses the effects of adjacent cell interference, power control and error correction on the performance of the base-to-mobile link of a multiple-cell CDMA cellular radio system. The performance measures considered in this analysis are the average bit-error-rate (BER) and the service reliability.

A statistical approach is used to estimate values for particular performance measures for a mobile anywhere in its cell. In the analysis, the system is assumed to employ coherent BPSK modulation and direct sequence spreading and the received signals are assumed to suffer Rayleigh fading, log-normal shadowing and frequency selective fading.

I. Introduction

As we march towards the era of personal communication networks (PCNs), there is a need to identify technologies that could provide the multiple access capabilities required to satisfy the large number of customers that PCNs are likely to attract. "Spread spectrum", or code division multiple access (CDMA), is a radio communications technique well known for its ability to allow multiple users to simultaneously use the same band and may be useful in providing effective high capacity systems in the future [1].

In a CDMA cellular system the effect of intra-cell interference and inter-cell interference varies depending on the position of the mobile in the cell and the power control and error correction schemes used. Generally error correction codes improve the performance of a system, but because such codes increase the bandwidth of the base band signal, in a CDMA system a reduction in the processing gain may be required to compensate. Thus the inclusion of error correcting codes in a CDMA system may not necessarily yield better performance. This paper analyses the effect of adjacent cell interference, power control and error correction on the performance of the base to mobile link of a CDMA cellular system.

Section II describes the system analysed in this paper and outlines the derivation of the probability density functions (PDFs) of the signal variables. Various system performance measures are defined in Section III. These measures are used to assess the capacity and quality of the CDMA system considered in this paper. The influence of adjacent cell interference is considered in Section IV. The effectiveness of power control and error correction in improving reception quality are outlined in Sections V and VI, respectively. As an example, short term average BER and service reliability results estimated for a mobile at the boundary of its cell are presented in Section VII. Finally, the conclusions drawn from this study are summarised in Section VIII.

II. System Description

In the analysis presented in this paper the quality of radio reception at a mobile in a CDMA cellular system is estimated and is averaged over the cell area. Figure 1 illustrates the type of scenario considered and shows three adjacent cells in a two-dimensional cell layout. (Note that up to 18 cochannel cells are considered in this paper).

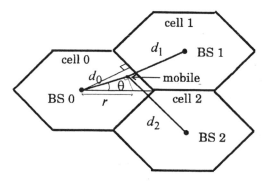

Figure 1: The base-to-mobile link and two adjacent interferers. (BS 0, BS 1, and BS 2 represent the base stations at the centre of the cells.)

The signals from the base station, received at the mobile are assumed to suffer Rayleigh fading, log-normal shadowing and frequency selective fading [2,3]. In some cases the wideband nature of CDMA channels may result in fading that is not Rayleigh distributed but is actually less severe. However in [4] it is reported that signals occupying a bandwidth of 1MHz can suffer Rayleigh-like fades exceeding 30dB, therefore for the purpose of this study it is reasonable to model the signal variability by a Rayleigh distribution and consider this as a "worst" case. The signals from the base stations are assumed to be asynchronous and to fade independently of each other. All the signals (including those intended for other users in the same cell) transmitted by a particular base station occupy the same frequency spectrum and propagate through the same multipath channel to arrive at a mobile. Therefore the CDMA signals from a particular base station, arriving at a given mobile, fade in unison. The frequency selectivity of the channel is modelled by the correlation bandwidth of the channel. The time delay corresponding to the correlation bandwidth is measured as a multiple of the chip period T_c [3].

All the users in the system under analysis are assumed to employ coherent BPSK modulation and direct-sequence (DS) spreading [5]. All the base stations and mobile antennas are assumed to be omnidirectional. The receiver employed in this CDMA system is assumed to reject all but one of the multipath components of the desired signal [3]. It is assumed that the desired user is connected to the base station in cell 0 and all the other base stations act as interferers. For simplicity it is assumed that there are $K+1$ users in cell 0 and K users in the other interfering cells (i.e. there are K interferers in each cell).

For a large number of users the combined interference can be approximated by a Gaussian random variable [3]. Having approximated the total interference, the momentary BER can be approximated by the complimentary Gauss probability integral of the signal-to-interference ratio (SIR) [5]. This approximation is known as the Gaussian approximation. When the number of users is small the Gaussian approximation underestimates the BER. However in this analysis we primarily investigate the influence of various system parameters on performance and are more concerned with identifying performance trends rather than precise values for performance measures (e.g. BER). However in most cases the results of this analysis are of greatest interest when the number of users is large and the system performance is verging on being unacceptable. In these situations the Gaussian approximation is appropriate. The SIR (voltage ratio) of the signal received by a mobile in the presence of interference from its own cell (cell 0) and L other cochannel CDMA cells (cells 1 - L) is given by [3,5]

$$SIR(\alpha_0,...\alpha_L) = \frac{\alpha_0}{\sqrt{a(K)\beta\delta\Omega_0\alpha_0^2 + b(K)\beta\delta(\Omega_1\alpha_1^2...+\Omega_L\alpha_L^2) + \frac{N_0}{2E}}}. \qquad ...(1)$$

In eqn. (1) $a(K)$ represents the self interference from the desired base station (in cell 0). This comprises of unwanted multipath components of the desired user's signal and the interference from signals intended for other users in the same cell. $b(K)$ represents the interference from users in the other cells. $N_0/2$ is the spectral density of the double sided additive white Gaussian noise (AWGN) at the receiver. E is the desired signal energy per bit. α_0 is the signal strength (voltage) received from base station 0, α_1 is the signal strength received from base station 1, etc. It is assumed that $\alpha_0, \alpha_1, ... \alpha_L$ are each Rayleigh distributed, and the means of the Rayleigh distributions are assumed to be log-normally distributed. The Ω_j (for $j = 1..L$) are defined as $\Omega_j = [d_j/d_0]^n$, where d_j is the distance between the jth user and the desired base station, d_0 is the distance between the user and the interfering base station in cell j, and n is the path loss exponent. By definition $\Omega_0 = [1/d_0]^n$. In eqn. (1) β is the voice activity factor and δ represents the interference reduction factor if power control is used. Assuming that $N_0/2E$ is small the momentary BER, $P(e|\alpha_0,...\alpha_L)$, is given by [5]

$$P(e|\alpha_0,...\alpha_L) \approx \Phi\left[\frac{\alpha_0}{\sqrt{a(K)\beta\delta\Omega_0\alpha_0^2 + b(K)\beta\delta(\Omega_1\alpha_1^2...+\Omega_L\alpha_L^2)}}\right], \qquad ...(2)$$

where

$$\Phi(x) = \frac{1}{\sqrt{2\pi}} \int_x^\infty \exp\left(\frac{y^2}{2}\right) dy. \qquad ...(3)$$

If $\gamma = \left[\Omega_1 \alpha_1{}^2 + \Omega_2 \alpha_2{}^2 + \ldots \Omega_L \alpha_L^2 \right] \alpha_0{}^{-2}$ is substituted into eqn. (2) then the expression for the momentary BER reduces to

$$P(e|\gamma) \approx \Phi\left[\frac{1}{\sqrt{a(K)\beta\delta\Omega_0 + b(K)\beta\delta\gamma}} \right].$$

...(4)

In order to determine the actual probability of error, the PDF of γ must first be determined. This can be obtained by evaluating the PDF of $\alpha_1{}^2 + \alpha_2{}^2 + \ldots \alpha_L^2$ (using Laplace transforms) and then evaluating the PDF of $\left[\alpha_1{}^2 + \alpha_2{}^2 + \ldots \alpha_L^2 \right] \alpha_0{}^{-2}$ (using Mellin convolution) [6]. The PDF of α_i is given by [2]

$$\text{PDF}(\alpha_i) = \frac{2\alpha_i}{\Gamma_i} \exp\left[-\frac{\alpha_i{}^2}{\Gamma_i} \right],$$

...(5)

where Γ_i is the mean-square value of α_i and is assumed to be log-normally distributed. If we substitute $\beta_i = \Omega_i \alpha_i^2$, then the PDF of β_i is given by

$$\text{PDF}_{\beta_i}(\beta_i) = \frac{1}{\Gamma_i \Omega_i} \exp\left[-\frac{\beta_i}{\Gamma_i \Omega_i} \right].$$

...(6)

The PDF of Γ_i is a log-normal distribution, namely [2]

$$\text{PDF}(\Gamma_i) = \frac{10}{\ln(10)\sqrt{2\pi}\sigma\Gamma_i} \exp\left[-\frac{\left(\frac{10}{\ln(10)} \ln\left(\frac{\pi}{4}\Gamma_i \right) - m_i \right)^2}{2\sigma^2} \right],$$

...(7)

where m_i is the area mean strength of the signal from the ith base station (in dBm) and σ represents the signal variability (in dB).

The PDF of γ has been determined to be

$$\text{PDF}_\gamma(\gamma) = \sum_{i=1}^{L} \frac{\gamma^{L-1}\Omega_i\Lambda_i}{(\gamma + \Omega_i\Lambda_i)^2} \prod_{\substack{j=1 \\ j \neq i}}^{L} \frac{1}{(\gamma + \Omega_j\Lambda_j)},$$

...(8)

where $\Lambda_1, \Lambda_2, \ldots$ and Λ_L are the ratios of the Rayleigh mean square values, namely $\Lambda_1 = \Gamma_1 / \Gamma_0$, $\Lambda_2 = \Gamma_2 / \Gamma_0$, $\ldots$ and $\Lambda_L = \Gamma_L / \Gamma_0$. The PDFs of Λ_i can be derived using Mellin convolution [6] and are given by

$$\text{PDF}_{\Lambda_i}(\Lambda_i) = \frac{10}{2\ln(10)\sqrt{\pi}\sigma\Lambda_i} \exp\left[-\left[\frac{\frac{10}{\ln(10)} \ln(\Lambda_i) - m_i + m_0}{2\sigma} \right]^2 \right].$$

...(9)

In the above equation the $m_i's$ are assumed to be functions of distance. It is assumed that $10^{m_i/10}$ is proportional to the inverse of the distance between the ith base station and the mobile raised to the power of the path loss exponent, i.e. $10^{m_i/10} \propto 1/d_i^n$ (refer Figure 1).

III. System performance measures

There are a number of possible measures of system performance that relate the received signal quality to the propagation conditions. Short-term average BER and service reliability are the most useful.

Short-Term Average BER :

The short-term average BER as a function of the number of users, $P_{av}(K)$, is defined to be the momentary BER averaged over the Rayleigh fading and is given by

$$P_{av}(K) = \int_0^\infty \Phi\left[\frac{1}{\sqrt{a(K)\beta\delta + b(K)\beta\delta\gamma}}\right] \times PDF_\gamma(\gamma)d\gamma.$$

...(10)

Service Reliability :

The service reliability, $P_{ser}(K)$, is defined as the percentage of time the momentary BER is below a maximum tolerable level, BER_{max}, i.e. $P_{ser}(K) = \text{Prob}(P(e|\gamma) < BER_{max})$. For K users per cell the service reliability can also be expressed as

$$P_{ser}(K) = \text{Prob}(SIR_K > SIR_{min}),$$

...(11)

where SIR_{min} is given by $\Phi(SIR_{min}) = BER_{max}$.

Given that

$$SIR = \frac{1}{\sqrt{a(K)\beta\delta + b(K)\beta\delta\gamma}},$$

...(12)

for a given value of SIR_{min} a corresponding value of γ_{max} can be found. Hence the service reliability is given by $P_{ser}(K) = \text{Prob}(\gamma < \gamma_{max})$ and can be estimated as

$$P_{ser}(K) = \int_0^\infty \cdots \int_0^\infty \int_0^{\gamma_{max}} PDF_\gamma(\gamma)PDF_{\Lambda_1}(\Lambda_1)\ldots PDF_{\Lambda_L}(\Lambda_L)d\gamma d\Lambda_1\ldots d\Lambda_L.$$

...(13)

IV. The effect of multiple interferers

In order to evaluate the quality of reception by a user in a cellular system, the interference from all the cochannel cells has to be considered. If the performance over the entire cell is to be estimated, then in a uniform propagation environment, due to the symmetry of the cells, the analysis can be limited to the smallest symmetrical unit of area. For a hexagonal layout this unit is a triangle, as shown in Figure 1, which occupies one twelfth of the cell area.

If the average performance in the entire cellular system over a long period of time is of interest, then the performance measures should be averaged over the log-normal shadowing and the cell area. In this paper, a processing gain of 128 and a voice activity factor of 0.5 are assumed. A BER threshold of 10^{-3} is used in the estimation of service reliability. The propagation conditions are assumed to be the same throughout the system with the variability of the log-normal shadowing, σ, equal to 6dB. The path loss exponent is assumed to have a value of 4.0. In the analysis presented in this section up to 18 cochannel interfering base stations are considered. These are assumed to be arranged in a regular hexagonal grid structure as shown in Figure 2.

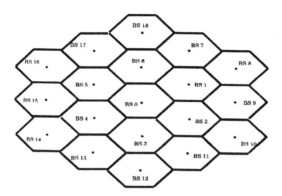

Figure 2: First and second rings of cochannel CDMA cells.

Figure 3 shows the effect of multiple cell interference on the average BER and the average service reliability. The average BER, $P_{av}(K)$, is obtained by averaging the momentary BER over the Rayleigh fading, the log-normal shadowing and the cell area, that is

$$P_{av}(K) = \int\limits_{\text{cell area } A} \int\limits_{0}^{\infty} \cdots \int\limits_{0}^{\infty} \int\limits_{0}^{\infty} \Phi\left[\frac{1}{\sqrt{a(K)\beta\delta + b(K)\beta\delta\gamma}}\right] \times PDF_\gamma(\gamma)\, d\gamma\, d\Lambda_1 \ldots d\Lambda_L\, dA.$$

...(15)

The average service reliability is estimated by averaging the service reliability at a point (eqn. (13)) over the entire cell area. In the calculation of the results in Figure 3 the "two interferers" are base stations 1 and 2 shown in Figure 2. The "six interferers" are the first ring of base stations (base stations 1 to 6). The "twelve interferers" are the first ring of interferers plus base stations 7 to 12, and the "eighteen interferers" are the complete first and second rings of base stations (base stations 1 to 18).

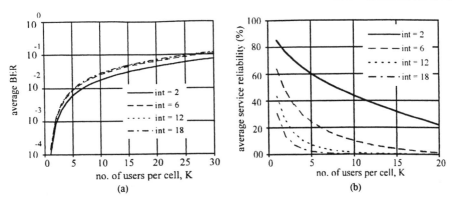

Figure 3: The effect of inter-cell interference on the performance of a CDMA system as a function of the number of users per cell with 2, 6, 12 and 18 interfering cells. A processing gain of 128, a path loss exponent of 4.0 and a σ value of 6dB were used in the evaluation.
(a) the average BER (evaluated using eqn. (15)).
(b) average service reliability (BER threshold for service reliability = 10^{-3}; evaluated by averaging the results obtained from eqn. (13) over the cell area).

The average BER curves indicate that the interferers outside the first ring of cells have negligible effect on the average BER. Interestingly however, the average service reliability curves show that the second ring of interferers have significant influence on the service reliability of the system. The difference in the sensitivity to the presence of a second ring of interferers can be attributed to the fact that the average BER is dependent largely on the average received interference power, while the average service reliability is more a function of the statistical variability of the received signals and the reliability threshold. Since the interference power from the second ring of cells is small compared to that from the first ring, the second ring has little effect on the average BER values. However it appears that the extra interference power from the second ring of cells tends to push the momentary BER from just below the threshold to just above the threshold, thus making the reception less reliable.

V. The effect of power control on the system

Power control is an essential requirement of a CDMA system [7]. Usually the power control is implemented by controlling the transmitted power from the base station so that it is proportional to the transmission distance (between the base station and the user), raised to the power of the path loss exponent [7]. Under this power control scheme the users at a short distance from the base station will be at a disadvantage in comparison to those further out. The signals intended for the users close to the desired base station can be swamped by the other (stronger) signals from the base station intended for the more distant users in the cell. Thus a service hole could exist close to the transmitter [7]. To avoid this situation a constant power could be transmitted to all users within a certain radius of the base station. Beyond this radius the power transmitted to a user would as before, be proportional to the transmission distance raised to the power of the path loss exponent. This power control algorithm can be expressed mathematically as

$$P(r) = P_R f(r),$$

...(16)

where $P(r)$ is the power transmitted to a user at distance r from the base station, P_R is a reference power level corresponding to the signal power transmitted to a user located at the cell boundary (where $r = R$), and

$$f(r) = \begin{cases} r^n & \text{for } r \geq aR, \\ (aR)^n & \text{for } r \leq aR, \end{cases}$$

...(17)

where a is the radius up to which the transmitted power is kept constant. Fig. 4 illustrates $P(r)$ for values of $a = 0, 0.75$ and 1.0 ($a = 1.0$ corresponds to no power control).

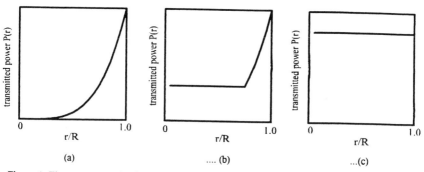

(a)   .... (b)   ...(c)

Figure 4: The power control schemes for (a) $a = 0$, (b) $a = 0.75$, and (c) $a = 1.0$ ($a = 1.0$ corresponds to no power control).

If the users are assumed to be uniformly distributed within a cell's area, then the PDF of their radial distribution is given by

$$p_r(r) = \frac{2r}{R^2}.$$

...(18)

Thus the power control factor, δ, is given by

$$\delta = \int_0^{aR} \left[\frac{r}{R}\right]^n \frac{2r}{R^2} dr + \int_{aR}^R \left[\frac{r}{R}\right]^n \frac{2r}{R^2} dr = \frac{2 + na^{n+2}}{n+2}.$$

...(19)

Figure 5 presents the average BER and average service reliability as a function of the number of users per cell for a range of power control factors a.

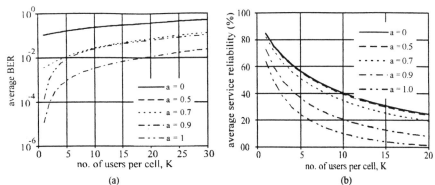

Figure 5: The effect of power control on the performance of a CDMA system as a function of number of users per cell for $a = 0, a = 0.5, a = 0.7, a = 0.9$ and $a = 1.0$ for six interfering cells. A processing gain of 128 and a σ value of 6dB have been assumed in the evaluation.
(a) the average BER (evaluated using eqn. (15) in conjunction with eqn. (19)).
(b) average service reliability (BER threshold for service reliability = 10^{-3}; evaluated by averaging the results obtained from eqn. (13) in conjunction with eqn. (19) over the cell area)).

In Figure 5, the curves obtained for $a = 0$ and $a = 0.5$ are almost the same because the difference in interference power for these two cases is almost negligible. The average BER curves indicate that a power control scheme with $a = 0.9$ gives the best performance, whereas the average service reliability curves indicate that a power control scheme with $a = 0$ gives the best performance. Once again it is the threshold requirement of service reliability that is responsible for the difference in the "optimal" power control factor. The power control factor that reduces the interference most effectively (i.e. $a = 0$) yields the best average service reliability. However this produces an area near the cell centre with very high BERs. These high BERs significantly raise the average BER.

VI. The effect of error correction

If we assume an (l,k) block code (i.e. for every k data bits l-k error correcting bits are added) capable of correcting all combinations of c and fewer errors, then the average BER, P_b, can be approximated by [8]

$$P_b \approx \frac{1}{l} \sum_{i=c+1}^{l} i \binom{l}{i} P_e^i (1 - P_e)^{l-i},$$

...(20)

where P_e is the corresponding BER in the absence of error correction. If the data is to be transmitted at a particular rate then as error correcting codes are added to the data the band-width occupied by the composite (base-band signal) code increases. If only a limited bandwidth is available for transmission, then the processing gain has to be reduced to compensate for the greater bandwidth of the base-band signal. Figure 6 shows the effects of error correction. A (23,12) Golay code for error correction is assumed, for which $c = 3$. In order to maintain the same transmission bandwidth, the error corrected data is assumed to have a processing gain of 66 and the uncorrected data a processing gain of 128.

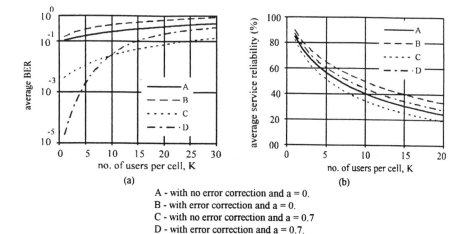

(a)

(b)

A - with no error correction and a = 0.
B - with error correction and a = 0.
C - with no error correction and a = 0.7
D - with error correction and a = 0.7.

Figure 6: The effect of error correction on the performance of a CDMA system as a function of the number of users per cell. A processing gain of 128 has been assumed for the system with no error correction and a processing gain of 66 has been assumed for the system with error correction. A σ value of 6dB is assumed.

(a) the average BER (evaluated using eqn. (15) in conjunction with eqn. (19) and eqn. (20)).

(b) average service reliability (BER threshold for service reliability = 10^{-3}; evaluated by averaging the results obtained from eqn. (13) in conjunction with eqn. (19) and eqn. (20) over the cell area)).

VII. General results

In Figure 7 short-term average BER and service reliability results are presented for systems with and without power control and error correction. All the results are calculated for a mobile at the boundary of its cell. The short-term average BER and service reliability are plotted as functions of the number of users per cell. A BER threshold of 10^{-3} and σ values of 6dB have been assumed in the estimation of service reliability.

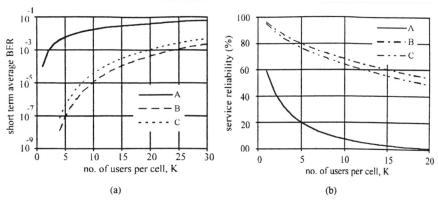

A - with no error correction and no power control
B - with error correction and a = 0.
C - with error correction and a = 0.7.

Figure 7: (a) Short term average BER (eqn. (10) was used to evaluate short term average BER), and (b) Service reliability (evaluated using eqn. (13) in conjunction with eqn. (19) and eqn. (20)) as a function of the number of users per cell. The results are estimated for a mobile at the boundary of the cell. A processing gain of 128 is assumed for the system with no error correction and a processing gain of 66 is assumed for the system with error correction in the evaluation. It is assumed that $\sigma = 6$dB and the BER threshold for reliable service is 10^{-3}.

VIII. Conclusions

From the results presented in this paper the following observations are made :

1. The interference from the second ring of interferers has negligible effect on the average BER, but has a significant effect on the service reliability.

2. Power control improves the performance of the CDMA system considered.

3. For a small number of users the error-corrected system considered in this paper performs better than the system with no error correction, but as the number of users increases, the situation reverses. (Note that both the systems occupy the same transmission bandwidth).

4. For a path-loss exponent of 4, the power control algorithm increases the capacity by approximately three-fold.

5. The combined effect of the power control and error correction considered in this paper was to increase the capacity by 200-400%, depending on the performance measure used in determining the maximum allowable number of users.

Acknowledgment

The authors wish to acknowledge the support provided by *Telecom Corporation of New Zealand Ltd*, and particularly thank Dr M V Clark for his useful comments on this work.

References

[1]   R L Pickholtz, L B Milstein and D L Schilling, "Spread spectrum for mobile communication," IEEE Trans. Vehic. Technol., pp. 313-322, May 1991.

[2]   K W Sowerby and A G Williamson "Outage probability calculations for a mobile radio systems with multiple interferers," *Electron. Lett.*, vol. 24, No. 24, Nov. 1988, pp. 1511-1513.

[3]   D E Borth and M B Pursley, "Analysis of direct-sequence spread-spectrum multiple-access communication over Rician fading channels," *IEEE Trans. Commun.*, vol. COM-27, pp. 1566-1577, Oct. 1979.

[4]   D L Schilling et. al, "Broadband CDMA for personal communications system," *IEEE Commun. Mag.*, pp. 86-93, Nov. 1991.

[5]   L B Milstein, T S Rappaport and R Barghouti, "Performance evaluation for cellular CDMA," *IEEE J Select. Areas Commun.*, vol 10, pp. 680-689, May 1992.

[6]   M D Springer, "The Algebra of Random Variables," Wiley, New York, 1979.

[7]   R R Gejji, "Forward-link-power control in CDMA cellular systems," *IEEE Trans. Veh. Tech.*, vol. 41, pp. 532-536, Nov. 1992.

[8]   D J Torrieri, "The information -bit error rate for block codes," *IEEE Trans. Commun.*, vol. COM-32, pp. 474-476, April. 1984.

6

The Design and Development of a Spread Spectrum

Digital Cordless Telephone for the Consumer Market

P. Stuckey McIntosh[*]

Abstract

The design and development of a low-cost spread spectrum consumer cordless telephone is described. The telephone operates in the 902-928 MHz band with a peak RF output power of 1 watt. The telephone combines a simple, direct-conversion receiver architecture with DSP techniques for spreading, coding, demodulation, correlation, and speech compression. Most DSP functions, all glue logic, the controlling state machine, and all circuitry for time division duplex (TDD) operation are implemented in a single ASIC. A telecom codec with ADPCM compression are utilized in the speech path, resulting in high (toll) speech quality. The transceiver subsystem, ASIC, and controlling firmware are described. Some performance benchmarks achieved by the product: digital security through three independent mechanisms; dropout-free operation; and 1 mile line-of-sight operation. The telephone has received FCC approval under Part 15.247 and is currently in production. Adaptive transmit power control and the multipath mitigation technique RSS (recombinant spread spectrum) are employed to give dropout-free performance while minimizing the possibility of interference to and from other systems. Some of the problems solved in the course of development are described, including digital-to-RF self-interference and isolation of sensitive, analog and audio circuits from the relatively high field-strength transmit signal.

Introduction

In 1985, new regulations were promulgated by the FCC that permit the use of direct sequence and frequency hopping spread spectrum in commercial products. The rules were amended and clarified in 1990. A consumer cordless telephone utilizing direct sequence spread spectrum is described here that complies with these new regulations and offers a higher level of performance than is attainable with analog-FM cordless phone technology.

System Design

The system design dictated a TDD radio link (economy) that was impervious to short-differential multipath propagation. In interior spaces, multipath delay spread, the time-of-arrival difference for multiple, discrete radio signals generally ranges from 10 to 150 nanoseconds. Additionally, rapid (120 Hz) multipath flutter caused by florescent lighting is common in interior spaces because of the cyclic ionization of conducting gases in the tubes. This creates a particularly onerous multipath problem. Direct sequence spread spectrum techniques provide resistance to multipath fading only if the multipath differential delay spread exceeds the spreading code chip width[1, 2]. If the differential delay spread is as low as 10 nanoseconds (common in interior spaces), the chip rate would need to be greater that 100 megachips per second to provide multipath resistance, giving an RF signal bandwidth of perhaps 70 MHz or so. This is more than the available bandwidth in the 902-928 MHz band.

_____

[*] Digital Wireless Corporation, 1442 Tullie Road, Atlanta, GA 30329, USA

RSS Modulation/Demodulation

A spread spectrum technique called Recombinant Spread Spectrum (RSS)[6], an extension of RZ-PSK[3, 4] was developed that provides extraordinary immunity to multipath fading and meets the requirements of FCC Part 15.247, the rule governing spread spectrum systems. The validity of RSS as an antimultipath technique can be demonstrated in both the time and frequency domains. This technique imparts multipath immunity in part by constantly changing a parameter (amplitude) of the RF signal. In RSS, information is carried in the phase of the signal, but the amplitude of the RF envelope is constantly changing. This effectively prevents a reflected ray, which is delayed in time, from ever perfectly cancelling out the earlier-arriving direct ray. Significant attenuation or flat fading can and does occur in multipath, but is worst when the handset is closest to the base station, which is coincidentally when signal strength (and SNR) is highest. So the flat-fading impact of multipath is greatest when signal strength is highest, and conversely, where signal strength is lowest (long range), multipath's flat-fading impact is diminished because the differential delay spread typically increases in proportion.

Multipath has adverse effects on both the phase and amplitude characteristics of digital modulations. These effects can be mitigated by proper techniques. In BPSK modulated RF carriers, two copies of the baseband information are transmitted, one in the upper and one in the lower sideband. All of the low frequency baseband information resides near the carrier. This low frequency information must be received in order to prevent excessive intersymbol interference (ISI) and maintain the receiver eye opening. Deep multipath fading is equivalent to putting a randomly variable notch filter in the communications channel. Multipath notching can obliterate much of its low frequency information when it occurs near the RF center frequency.

When bipolar RZ coding (Fig. 1) is applied to the data stream prior to modulation, the RF bandwidth of the transmitted signal is doubled. The low frequency information is more widely dispersed. More importantly, the signal has a constantly changing, sinusoidal amplitude envelope. In order for signal cancellation to occur in a multipath propagation situation, two or more transmitted rays with net equal amplitude and opposite phase must arrive at the receive antenna. There is inevitably a short time delay between the arriving signals, and if some parameter of the signal is constantly changing, the reflected, delayed signals can never "catch up" in time with the direct ray. This limits or minimizes the amount of damage that multipath can do to the signal. This principal of signal dynamics works in concert with the additional frequency dispersion and redundancy of direct sequence spread spectrum to achieve an extraordinary degree of multipath immunity without the use of diversity antennas.

The upper bound on differential delay spread tolerance of this system is reached when the delay spread equals the chip rate[3, 4]. This is not a concern for this system, because path length difference would need to be nearly ½ mile before this bound is reached. This sort of differential delay is simply not encountered in the indoor residential/commercial environment owing to the magnitude of typical physical dimensions and the moderate but steady attenuation of UHF radio signals as they pass through multiple thicknesses of typical interior building materials.

In addition to the signal-redundant, bipolar RZ coding, chip-differential demodulation is employed so that the phase of the information signal is always decoded relative to the phase of the immediately preceding chip. Chip-level differential demodulation is employed rather than bit-level differential encoding because of frequency tolerance considerations. The transmitter-receiver LO offset tolerance in a system employing differential demodulation is inversely proportional to the length of the delay: the longer the differential delay, the less transmitter/receiver frequency offset allowed. In the system discussed here, the LO offset tolerance is about ± 10 kHz. This is easily achieved with an economical, standard AT-cut crystal without temperature compensation. With bit-level differential encoding, LO offset tolerance would have been reduced to about 1.5 kHz, necessitating the use of a more expensive TCXO reference for both handset and base.

A transmitter with differential encoder and receiver with bandpass differential demodulation are shown in Fig. 2. This transmitter structure is straightforward to implement, but the receiver is not. The long delay for differential demodulation requires a SAW filter. Fig. 3 shows the same receiver function implemented with DSP. Because of the sampling and processing rate requirement, general-purpose DSPs are too slow for this application. A hardware DSP implementation was developed that performs this function. The design was originally implemented in Xilinx arrays, then transferred to a standard-cell ASIC. In every digital radio design, there is a partition between the RF section and the digital section. In this receiver design, we have moved the RF/digital partition closer to the antenna and included band shaping, demodulation and AGC in the digital portion. We look forward to the day when we can implement the receiver front end in DSP (not really; we are RF guys at heart).

There were three objectives in the transceiver design: to implement the entire transceiver with one LO, to translate the receiver bandpass filter into its lowpass equivalent (eliminating SAWs), and to implement some of the filtering and all of the demodulation and despreading (correlation) functions in DSP. A single-LO, DSP-intensive transceiver implies a direct conversion receiver. Direct conversion receivers typically have moderate RF gain and high baseband gain. This can lead to severe DC settling-time problems, especially in a TDD receiver where the refractory period allowed for transmit-to-receive switching is very short.

In this design, the DC settling-time problem was solved by AC-coupling the I and Q baseband signals at a relatively high cutoff frequency. This introduces significant ISI. In this system, we used intentional, controlled ISI so that the transmitted waveform, which is 3-level (bipolar) RZ-coded, low-pass-filtered arrives at the demodulator as Manchester-coded baseband, another code that can be differentially demodulated[5]. One way of analyzing this transformation of this is to view bipolar RZ code as "preintegrated" Manchester code. When the bipolar RZ is then high-pass filtered (differentiated), Manchester results. We call this technique frequency-domain transcoding, or FDT. It is very effective in achieving the fast DC settling time required in TDD digital radio.

ASIC

Pulling together the logic, clock recovery, synchronization and FIFO structures for implementing the Time Division Duplex (TDD) engine was relatively straightforward. An ordinary telecom voice codec is used to encode the voice and a dedicated DSP chip performs ADPCM compression to minimize RF channel bit rate. Supervisory data with parity is passed between handset and base micros constantly to provide control and to monitor voice channel quality or BER. Supervisory bits are distributed evenly throughout the voice packet in order to increase the probability of detecting burst errors. A simple Command Retransmission And Parity error-checking mechanism minimizes spurious operation and provides a method for calculating BER, one mechanism that is used to infer voice link quality.

Adaptive Power Control

An adaptive power control mechanism sets transmit power to one of four levels: 1 mW, 10 mW, 100 mW or 1W. Power is adjusted under far-end, closed-loop control. The minimum RF transmit power level required for reliable communications is always used. It is assumed that the handset will be used very near the base most of the time. Adaptive power control then accomplishes three things: 1) it minimizes the radius of the "zone of interference" created by the transmitter in the 902-928 MHz band, thus reducing the possibility of interference to similar and dissimilar systems alike (electromagnetiquette); 2) it extends battery life by conserving energy during talk time; and 3) it reduces the possibility of EMI disturbances in nearby receivers, audio systems, and other unintentional RF detectors. With a transmitter operating at one watt in a TDD mode, the TDD process superimposes a 250 Hz, square-wave AM envelope onto the transmitted signal. This 250 Hz signal, like radar, will find its way into audio equipment such as CD players and stereos, many of which do not have effective shielding against strong UHF fields.

Self Interference

In the course of testing early prototypes, several unanticipated problems were uncovered. First, receiver range was much shorter when the self-contained antenna was used in place of a remote antenna. It was determined that this was caused by broadband, relatively high field-strength signals being generated by the various HCMOS digital circuits. Since these circuits are all clocked at frequencies below 10 MHz, this means that our receiver was being jammed by the 90th and higher harmonics -- hard to believe, but true. This is a testament to the switching speeds that are now being attained by in the HCMOS digital logic.

Additionally, the TDD envelope of the 1-watt transmitter was rectified and picked up by any exposed audio circuits. Telecom codecs have high-gain audio opamps in them. The conventional, telephone-type electret microphone has an internal FET buffer that rectifies or detects the RF transmit envelope (both TDD rate and SS chip rate). These problems were solved by judiciously placed (well grounded) bypass capacitors (18 pF was found to be self resonant in the 0805 size at 915 MHz), along with de-Qing resistors of appropriate values in shunt/series arrangement on all transceiver I/O leads. For high-current lines, ferrite bead chips were used for this de-Qing.

Voice Channel Security

Communications security in this product is high for a consumer item. First, voice is digitally coded with a standard 8-bit telecom codec. Then, non-standard rate ADPCM speech compression is used. This is followed by convolution with a 31-bit PN sequence, followed by spreading. The 31-bit code and spreading together provide a high degree of randomness to the RF signal so that there are no significant discrete components. The regime of the 31-bit code is changed synchronously at the base unit and handset once every 768 packets (about once every three seconds). This change is made according to an algorithm that is seeded by the 24-bit unit ID that is unique to the handset and base. This assures that the scrambling pattern is different for each unit, and that whatever you say while using this product will remain a mystery to your neighbors, your mother, the IRS, and others.

Clear Channel Selection Algorithm

The RF bandwidth of the product is approximately 1.4 MHz. This allows about 20 channels in the 902-928 MHz range. Each handset-base pair picks three of the available 20 channels for its primary channels. Scanning three channels allows for frequency redundancy should one or more of the channels be jammed by interference. If interference is encountered on one of these three channels, it is dropped from the scanning sequence and another, again chosen by an algorithm seeded by the unique 24-bit unit ID, will be substituted for it. This effects a prearrangement between handset and base regarding channel assignment keeps handset and base in sync in the event of jamming. The ID-seeding randomizes the channel assignment process to minimize the probability of irresolvable interference between like units.

Conclusion

The significant objectives of the design, 1-mile line-of-sight range, corded telephone sound quality, low cost, and a minimum number of adjustments in production test were achieved. Additional cost savings and better manufacturing economy can probably be achieved by higher level integration of RF circuits. Additional systems and products are envisioned that will utilize many of the principals, practices, and techniques developed for this product.

REFERENCES

[1]   A. J. Viterbi, "The orthogonal-random waveform dichotomy for digital mobile personal communication," *IEEE Pers. Comm.,* vol. 1, pp. 18-23, First qtr. 1994.

[2]   R. C. Dixon, *Spread Spectrum Systems.* New York: Wiley, 1984, ch. 7.

[3]   S. Yoshida, F. Ikegami, S. Ariyavisitakul and T. Takeuchi, "Delay-spread-resistant modulation techniques for mobile radio," *Proc. IEEE GLOBECOM,* Tokyo, Japan, Nov. 1987, pp. 811-817.

[4]   S. Ariyavisitakul, S. Yoshida, F. Ikegami, K. Tanaka and T. Takeuchi, "A power-efficient linear digital modulator and its application to an anti-multipath modulation PSK-RZ scheme," *IEEE VTC'87*, Tampa, Florida, June 1987, pp. 1-6.

[5]   S. Yoshida and F. Ikegami, "Anti-multipath modulation technique - manchester-coded PSK," *IEEE ICC'87,* Seattle, WA, 1987, pp. 1371-1375.

[6]   US patent #4,862,478 and other patents pending.

Stuckey McIntosh (M '80) received the bachelor's degree in electrical engineering from the Georgia Institute of Technology in 1980. He analyzed Soviet surface-to-air missile systems for Georgia Tech Research Institute (GTRI). He designed early equipment for transmitting news wire services over the video vertical blanking interval (VBI) via satellite. As a member of the cable communications group at Scientific-Atlanta, he developed CATV headend gear, set-top converter systems, and parts of high-speed CATV modems. He co-founded LanTel, a manufacturer of CATV modems for voice and low-speed data. He holds one spread spectrum patent and has others pending. In 1987, he co-founded and became Chairman of Gambatte Digital Wireless (now Digital Wireless Corporation), Atlanta, Georgia, which develops and manufactures spread spectrum products for OEMs. Since 1987, his focus has been on the commercialization of low-cost, spread spectrum voice and data communications products and systems for the ISM bands. He contributed to the FCC Part 15 rules that govern the commercial spread spectrum industry.

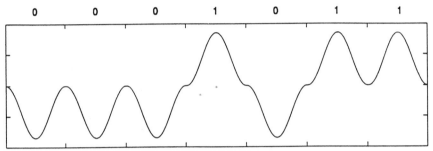

Fig. 1. Bipolar-RZ encoded data

104

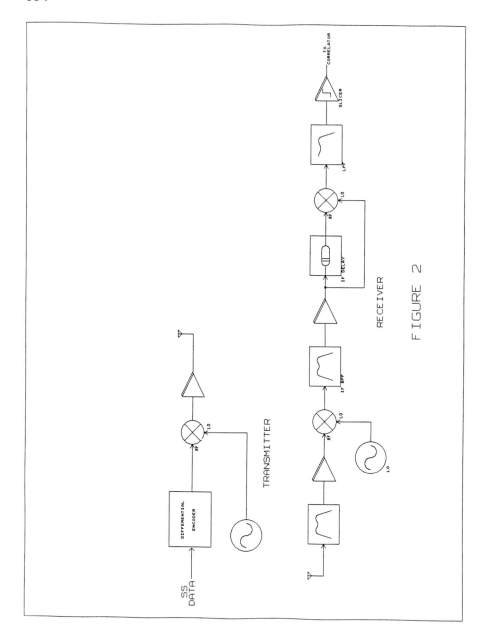

TRANSMITTER

RECEIVER

FIGURE 2

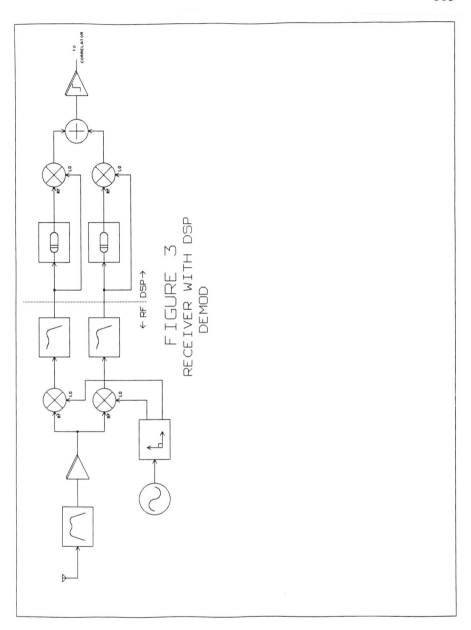

FIGURE 3
RECEIVER WITH DSP
DEMOD

7

A Principled Framework for Narrowband Mobile Digital Communications

Michael P. Fitz
email: mpfitz@ecn.purdue.edu
Phone: (317) 4940592

James P. Seymour
email: seymour@ecn.purdue.edu
Phone: (317) 494-3489

School of Electrical Engineering
Purdue University
West Lafayette, IN 47907-1285
Fax (317)-494-0592

Abstract

This paper developes a model-based framework for narrowband communications. First, we derive the optimum (MAP) demodulation algorithm for M-QAM signaling in the frequency flat time-varying Rayleigh fading channel. This recursive structure allows for symbol-by-symbol data decisions to be made in an efficient manner. The complexity problem inherent to optimal detection schemes is addressed and approximations to optimal detection are considered which reduce complexity to practical levels yet still provide near-optimal performance levels. Reduced complexity algorithms are obtained through the use of decision feedback, thresholding and a novel state partitioning which allow for higher order modulations such as 16-QAM and 64-QAM to be implemented. The algorithms presented exhibit favorable advantages over pilot symbol assisted modulation techniques in performance, bandwidth efficiency and the necessary decoding delay. This principled approach to narrowband mobile digital communications provides a unified framework which incorporates diversity combining techniques, array signal processing, forward error control decoding and optimum demodulation in a single demodulation architecture. Performance characterizations of the algorithms presented are obtained via Monte Carlo simulation and the results show that near-optimal BEP levels are achieved.

1. Introduction

Interest in mobile communications applications (e.g., cellular telephone and personal communications) has rapidly increased in recent years which has resulted in a scarcity of spectrum available for such applications. Consequently, there is a need for high performance and bandwidth efficient transmission techniques for use in time-varying fading channels. This paper considers narrowband land mobile digital communications. Narrowband communications in the context of this paper refers to digital communications which have instantaneous bandwidths which are much smaller than the correlation bandwidth of the channel. Note that CDMA systems which use slow frequency hopping are often narrowband according to this definition while utilizing a large transmitted spectrum. A major advantage of narrowband communications is that the small bandwidths make low power transceiver implementations more feasible.

Narrowband digital communications is accurately modeled as imposing a time-varying

108

multiplicative distortion (MD) process upon the transmitted signal. This MD can be rapidly varying (e.g., see Fig. 1) and high performance digital communication on this type of a channel can be a significant challenge. Bandwidth efficient digital communications requires the use of

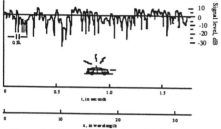

Figure 1. The envelope of the multiplicative distortion typical of narrowband radio channels.

large signal constellations (e.g., 16-QAM). The most significant impairment to demodulation with large constellations is producing an estimate of the time-varying MD for coherent demodulation. Common techniques for estimating the MD process are transmitted reference schemes such as pilot symbol assisted modulation (PSAM) [1, 2] and tone calibration techniques (TCT) [3-6].

Transmitted reference techniques are popular in mobile communications environments because they are simple to implement, they reduce the bit error probability (BEP) floor inherent to differential detection in time-varying fading, and they produce near coherent performance with large signal constellations. Unfortunately, additional power and/or bandwidth is necessary with such techniques for the generation of the reference signal. For example, PSAM techniques require the transmission of known pilot symbols throughout the information bearing symbols in order to produce the demodulation reference signal. The maximum spacing between pilot symbols is restricted by the Doppler spread introduced by the channel. For large Doppler spreads this can significantly reduce the effective data transmission rate, causing PSAM implementations to have reduced bandwidth efficiencies.

The transmitted reference schemes are also less efficient since the MD estimation is only based on the information in the transmitted reference. Using the information contained in the modulated, yet information bearing portions of the signal can significantly improve the performance. More efficient demodulation architectures can be obtained if the estimation algorithm is based on probabilistic characterizations of the entire received signal. This paper investigates MAP symbol-by-symbol detection algorithms for M-QAM signaling assuming the transmitted signal is subjected to frequency non-selective time-varying fading and corrupted by AWGN. This work is based on the classic K-lag recursive estimation algorithm of Abend and Fritchman [7]. While the optimal algorithm is computationally intractable, several approximations using decision feedback and other complexity reduction techniques provide near ideal performance in a computationally feasible form.

MAP symbol-by-symbol demodulation provides several performance advantages. These algorithms provide better bit error probability (BEP) performance than standard PSAM by using both reference and modulated symbols in demodulation. Although the MAP demodulation algorithms also require the transmission of pilot symbols in order to avoid cycle slipping, the pilot symbol spacing is not restricted by the Doppler spread of the channel as with PSAM. Thus, higher effective data rates and increased spectral efficiencies are possible without degraded performance. Also, the decoding delay of MAP demodulation is typically much shorter than that required for equivalent performance with PSAM processing. The MAP estimation structures are also easily generalized to provide optimum diversity combining (including interleaving and coding). Finally, symbol-by-symbol demodulation produces the metrics needed for soft decision decoding (i.e., posterior symbol probabilities) in contrast to optimum sequence demodulation methods. These advantages of the MAP symbol-by-symbol demodulators are quantified and the general characteristics of these algorithms are explored in this paper.

The paper is organized such that Section 2 presents the analytical models for narrowband digital communication and Sections 3 and 4 overview optimum and recursive structures for demodulation. Section 5 presents some of the complexity reduction schemes making these ideas feasible for large modulations. Section 6 and 7 discuss how the processing of multiple antennas and the decoding of forward error control codes can be incorporated into the model based demodulation framework. Section 8 concludes.

2. Analytical Models

2.1. Transmitted Signal

A block diagram of the system structure considered in this work is shown in Fig. 2. We consider linear modulations, so the transmitted signal has the form

$$z(t) = \sum_i d_i u(t - iT) \tag{1}$$

where $u(t)$ represents a square root unit energy Nyquist pulse shape[1] and T is the symbol time. The transmitted modulation symbols are complex valued and taken from an alphabet Ω_d of size M. The d_i are normalized for the specific modulation type so that $E\{|d_i|^2\} = 1$. The information bearing symbols are assumed to be i.i.d. random variables since interleaving is a common practice in fading channel communications. This paper considers MPSK and M-QAM signals.

Since the signal sets considered in this paper all exhibit rotational symmetry and produce a potential phase ambiguity for demodulation, we use PSAM. The known pilot symbols of PSAM are inserted periodically in the transmitted symbols to aid in deriving a channel state estimate, to resolve phase ambiguities, and prevent cycle slipping at the demodulator. The value of the known pilot symbols is typically randomized [8] for spectral shaping and interference suppression, and we assume that the receiver has knowledge of the values of the pilot symbols. The pilot symbols are assumed to be evenly spaced and the pilot spacing is denoted by N_p.

---

[1] $R_u(nT) = \int_{-\infty}^{\infty} u(t)u^*(t - nT)dt = \delta_n$

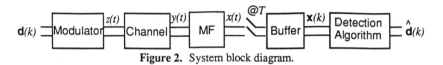

Figure 2. System block diagram.

2.2. Channel Model

A channel model used for the development of algorithms in this paper is the frequency nonselective, time-varying, isotropic scattering, Rayleigh fading channel. With this channel the input to the demodulator is given by

$$y(t) = c(t)z(t) + v(t) \tag{2}$$

where $c(t)$ is a complex Gaussian multiplicative distortion (MD) and $v(t)$ is a complex AWGN process with one-sided spectral density N_0. The isotropic Rayleigh scattering assumption implies that $c(t)$ is characterized by the Fourier transform pair

$$R_c(\tau) = E\{c(t)c^*(t-\tau)\} = \overline{E}_s J_0\left(\frac{2\pi f_D \tau}{T}\right) \tag{3a}$$

$$S_c(f) = \frac{\overline{E}_s T}{\pi f_D}\left(1 - \frac{(fT)^2}{f_D^2}\right)^{-\frac{1}{2}} \qquad |f| < \frac{f_D}{T} \tag{3b}$$

where f_D is the normalized Doppler spread and $\overline{E}_s$ is the average received symbol energy.

2.3. Demodulator

The first step in the demodulation process is to pass the received signal (2) through a filter matched to the pulse shape $u(t)$. This paper assumes that the pulse shape is narrow compared to $1/f_D$ (slow fading assumption) so that the MD is approximately constant for a pulse duration and the ISI produced by the MD is negligible. The sampled output of the matched filter (MF) is then

$$x_i = d_i \int_{-\infty}^{\infty} c(iT - \lambda)|u(-\lambda)|^2 d\lambda + v_i' \approx d_i c_i + v_i' \tag{4}$$

where c_i is a discrete time Gaussian MD process with $E\{c_i c_{i-l}^*\} \approx R_c(lT)$ and v_i' is a white Gaussian noise sequence with variance N_0.

3. Overview of Optimum Demodulation in Fading

First we will consider the optimum detection of uncoded modulation symbols. The algorithm developed here roughly parallels that of [9]. Define the following vectors

$$\begin{aligned}
\mathbf{d}(k) &= \left[d_k, d_{k-1}, \ldots, d_1\right]^T & \mathbf{d}_n(k) &= \left[d_k, \ldots, d_{k-n}\right]^T \\
\mathbf{x}(k) &= \left[x_k, x_{k-1}, \ldots, x_1\right]^T & \mathbf{x}_n(k) &= \left[x_k, \ldots, x_{k-n}\right]^T \\
\mathbf{c}(k) &= \left[c_k, c_{k-1}, \ldots, c_1\right]^T & \mathbf{c}_n(k) &= \left[c_k, \ldots, c_{k-n}\right]^T.
\end{aligned} \tag{5}$$

The minimum symbol error probability detection algorithm is given by

$$\hat{d}_{k-K} = \arg \max_{d_{k-K} \in \Omega_d} p\big(d_{k-K}|\mathbf{x}(k)\big). \tag{6}$$

This probability mass function (pmf) is computed by total probability and the decision rule becomes

$$\hat{d}_{k-K} = \arg \max_{d_{k-K} \in \Omega_d} \sum_{\mathbf{d}(k-K-1),\mathbf{d}_{K-1}(k)} p\big(\mathbf{d}(k)|\mathbf{x}(k)\big). \tag{7}$$

Using Bayes theorem, the posterior density can then be written as[1]

$$p\big(\mathbf{d}(k)|\mathbf{x}(k)\big) = \frac{f\big(\mathbf{x}(k)|\mathbf{d}(k)\big)p\big(\mathbf{d}(k)\big)}{f(\mathbf{x}(k))} \tag{8}$$

and the optimum decision rule is then

$$\hat{d}_{k-K} = \arg \max_{d_{k-K} \in \Omega_d} \frac{\sum_{\mathbf{d}(k-K-1),\mathbf{d}_{K-1}(k)} f\big(\mathbf{x}(k)|\mathbf{d}(k)\big)p\big(\mathbf{d}(k)\big)}{\sum_{\mathbf{d}(k)} f\big(\mathbf{x}(k)|\mathbf{d}(k)\big)p\big(\mathbf{d}(k)\big)}. \tag{9}$$

Note in (8), $f(\mathbf{x}(k))$ is only a normalizing constant (see (9)) and is not a function of $\mathbf{d}(k)$ so in theory it can be ignored in an optimal detection scheme. Also note that $p(\mathbf{d}(k))$ is not a constant due to the use of PSAM. Thus, $f\big(\mathbf{x}(k)|\mathbf{d}(k)\big)p(\mathbf{d}(k))$ represents a sufficient statistic for making the decisions $\hat{d}_{k-K}$. From (7) and (8) note there are M^k sufficient statistics at the kth time sample.

The Gaussian nature of the MD and the noise make the calculation of the pmf in (8) straightforward. The second order conditional moments of $x(k)$ are given by

$$E\big[\mathbf{x}(k)|\mathbf{d}(k)\big] = 0 \qquad \mathbf{C}_{\mathbf{x}}(k;\mathbf{d}(k)) = E\big[\mathbf{x}(k)\mathbf{x}(k)^H|\mathbf{d}(k)\big] = \mathbf{\Lambda}_{\mathbf{d}}(k)\mathbf{C}_{\mathbf{c}}(k)\mathbf{\Lambda}_{\mathbf{d}}(k)^H + \mathbf{I}_k N_0 \tag{10}$$

where $\mathbf{C}_{\mathbf{c}}(k) = E\big[\mathbf{c}(k)\mathbf{c}(k)^H\big]$, $\mathbf{I}_k$ is the $k \times k$ identity matrix and $\mathbf{\Lambda}_{\mathbf{d}}(k) = diag\big(d_k, d_{k-1},..., d_1\big)$. The optimal K-lag smoothing detector can now be implemented, noting that the conditional density in (8) is given by

$$f\big(\mathbf{x}(k)|\mathbf{d}(k)\big) = \frac{1}{\pi^k \det\big(\mathbf{C}_{\mathbf{x}}(k;\mathbf{d}(k))\big)} \exp\Big(-\mathbf{x}^H(k)\big(\mathbf{C}_{\mathbf{x}}(k;\mathbf{d}(k))\big)^{-1}\mathbf{x}(k)\Big). \tag{11}$$

The mass function $p(\mathbf{d}(k))$ takes on a constant value or equals zero depending on whether or not the sequence $\mathbf{d}(k)$ contains the correct values at the pilot symbol positions. The exponential growth in complexity of this detection scheme is evident since there are M^k sufficient statistics that must be calculated at the kth stage, all of which require the $k \times k$ matrix inverse in (11). This detection scheme is only practical for short data packets like those in a slow frequency hopped communication system. A recursive form of this algorithm is more practical for continuous transmission and also eliminates the need for the large matrix inverse in (11).

---

[1] $p(\cdot)$ and $f(\cdot)$ indicate a pmf and a probability density function (pdf), respectively.

4. Recursive Demodulation Structures

The computation for $f(x(k)|d(k))p(d(k))$ can be spread over all k symbols with a recursive algorithm similar in form to that found in [7]. Note that $f(x(k-1)|d(k-1))p(d(k-1))$ is now the information state and represents the kth stage sufficient statistic for recursive optimal demodulation (c.f., [10] for a discussion of sufficient statistics and information states in recursive estimation). Assuming that $f(x(k-1)|d(k-1))p(d(k-1))$ is known from the previous update and that x_k has been observed, then $f(x(k)|d(k))p(d(k))$ can be written as

$$
\begin{aligned}
f\big(x(k)|d(k)\big)p\big(d(k)\big) &= f\Big(x(k-1), x_k \Big| d(k-1), d_k\Big) p(d(k-1)) p\big(d_k\big) \\
&= f\Big(x_k \big| x(k-1), d(k)\Big) p\big(d_k\big) f\Big(x(k-1)\big|d(k-1), d_k\Big) p(d(k-1)).
\end{aligned}
\tag{12}
$$

Note that the samples in $x(k-1)$ are not a function of d_k and thus knowledge of d_k does not affect the statistics of $x(k-1)$. Therefore,

$$
f\big(x(k)|d(k)\big)p\big(d(k)\big) = f\Big(x_k \big| x(k-1), d(k)\Big) p\big(d_k\big)\Big[f\big(x(k-1)|d(k-1)\big) p(d(k-1))\Big]
\tag{13}
$$

provides the desired recursion formula for $f(x(k)|d(k))p(d(k))$ (the updated sufficient statistic) because the quantity in brackets is the sufficient statistic from the previous update. Since $f(x(k)|d(k))$ is Gaussian (see (11)), well known results for conditional distributions of Gaussian random variables yield

$$
f\Big(x_k \big| x(k-1), d(k)\Big) = \frac{1}{\pi\, \varepsilon(k-1; d(k))} \exp\left(-\frac{\big|x_k - a(k-1; d(k))^T x(k-1)\big|^2}{\varepsilon(k-1; d(k))} \right)
\tag{14}
$$

where $a(k-1; d(k))$ is the vector of conditional forward linear prediction coefficients and $\varepsilon(k-1; d(k))$ is the associated conditional forward prediction error variance [11]. The conditional mean in (14) is the output of a length k linear prediction filter whose coefficients are dependent on the sequence, $d(k)$, being considered. This relationship between linear prediction filter theory and recursive MAP detection is an interesting characteristic of this recursive algorithm for flat Rayleigh fading. By standard results for conditional moments of Gaussian vectors, the forward prediction filter coefficients are given as

$$
a(k-1; d(k))^T = d_k c_1(k-1)^H \Lambda_d(k-1)^H C_x(k-1; d(k-1))^{-1}
\tag{15}
$$

where

$$
c_1(k-1)^H = E\Big[c_k c(k-1)^H\Big].
\tag{16}
$$

The prediction error variance is similarly given as

$$
\varepsilon(k-1; d(k)) = N_0 + \big|d_k\big|^2 \Big[\overline{E}_s - c_1(k-1)^H \Lambda_d(k-1)^H C_x(k-1; d(k-1))^{-1} \Lambda_d(k-1)c_1(k-1)\Big]. \tag{17}
$$

Equation (14) provides a simple method for updating the information state for optimal symbol-by-symbol demodulation. The final data estimate is still obtained by using (13) and (8) in (7). However, note that (13) produces M sufficient statistics at time k for every sufficient statistic at time k-1 (one for each possible d_k) and the recursive algorithm still has complexity $O(kM^k)$. While the recursive implementation is computationally more efficient, it does nothing toward reducing the exponential complexity with time. This is in direct contrast to the demodulation scheme of [7] which produces a fixed complexity, $O(M^K)$, per update. The reason for this difference is that in [7] the channel was assumed known, while here we only assume a statistical description of the channel. This recursive algorithm is not only estimating the data but also (although not explicitly) estimating the channel and information about the channel is contained in the entire received sequence. Consequently, the optimal recursive algorithm uses the entire received sequence and has a recursion complexity which grows exponentially with the observation window.

5. Complexity Reduction Techniques

The complexity of the K-lag detector is significantly reduced by using four techniques. These techniques use feedback of the final decisions in the algorithm recursion, a finite sliding window observation vector for the computation of the innovation term (i.e., (14)), a statistic pruning technique in the recursive algorithm, and decision feedback within a reduced state partition of the transmitted symbols. The remainder of this section overviews these complexity reduction techniques and more details can be found in [12].

This family of optimal demodulators are formulated with a K-lag architecture. These K-lag decisions can be fed back into the recursive algorithm to simplify the recursions. The use of decision feedback is equivalent to assuming that d_{k-K} is deterministic given $\mathbf{x}(k)$ (i.e., d_{k-K} is $\mathbf{x}(k)$ measurable). The demodulator now recursively computes $p(\mathbf{d}_K(k)|\mathbf{x}(k))$ to produce a recursive estimator of the form

$$\hat{d}_{k-K} = \arg \max_{d_{k-K} \in \Omega_d} \sum_{\mathbf{d}_{K-1}(k)} p(\mathbf{d}_K(k)|\mathbf{x}(k)). \tag{18}$$

Note in (18) the conditional mass functions are also condition on the past decisions, but this conditioning is suppressed both to simplify notation and because these decisions are functions of $\mathbf{x}(k)$. The information state for this recursive algorithm now has size M^{K+1} and the algorithm complexity is driven by the update of this information state.

The use of a finite sliding observation window for the innovation term does not significantly degrade the BEP performance since a long observation window is primarily needed for providing very accurate estimates of the MD process. Fortunately $\lim_{\tau \to \infty} R_c(\tau) = 0$ (i.e., a finite coherence time) and the quality of the channel estimates necessary to achieve near optimum performance is less than two orders of magnitude better than the matched filter output quality. Thus, a reasonable value for the window size provides the desired performance. The finite observation window approximation is equivalent to the following approximation

$$R_c(mT) = 0 \qquad \forall \, |m| > N. \tag{19}$$

This approximation leads to a finite length prediction filter for the calculation of the conditional mean for the innovation term (14). The performance of this demodulation algorithm for BPSK modulation as function of K and N is seen in Fig. 3 and 4, respectively. Note the ideal curve in Fig. 3 correspond to ideal demodulation with N_p=5.

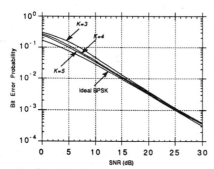

Figure 3. Simulated BPSK BEP performance of model-based demodulation for different smoothing lengths K. N_p=5 and f_D=.02.

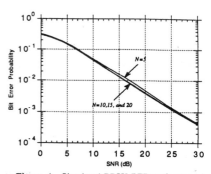

Figure 4. Simulated BPSK BEP performance of model based demodulation for different N. N_p=5 and f_D=.02.

The statistic pruning is accomplished by realizing sufficient statistics produced in the recursive algorithm are approximately equal to the posterior mass of the most recently transmitted symbols (i.e., $\approx p\big(\mathbf{d}_{K-1}(k)|\mathbf{x}(k)\big)$). These statistics can be used to decide which states are "unimportant" (i.e., highly unlikely) and remove those from the recursion. An adaptive complexity reduction scheme compares the statistics for each $\mathbf{d}_{K-1}(k)$ in the recursion to a threshold p_t and removes states from the recursion if their corresponding statistics are less than the threshold. This significantly reduces the number of statistics in the recursion especially in fading channel applications.

This complexity reduction scheme has several advantages in fading channels. First of all, the average SNR in fading channel communications is usually high. This implies that a majority of the decisions are very reliable and it will be easy to eliminate states by examining the posterior mass function. If multiple states are retained in a threshold scheme such as this, then one of the symbols in $\mathbf{d}_{K-1}(k)$ is typically experiencing a deep fade. Since performance in fading is generally dominated by the performance during deep fades, the main advantage of thresholding is to yield high complexity only at the times when performance is critical (low SNR) while keeping complexity low during less critical times (high SNR). The threshold value, p_t, determines both the performance and the complexity levels. Larger values of p_t produce greater complexity reduction while potentially reducing performance by eliminating useful information. On the other hand, smaller values of p_t provide potentially better performance levels at the cost of higher complexity. Fig. 5 and 6 show the bit error probability of this reduced complexity demodulator for 16-QAM and average number of states saved in the algorithm recursion, respectively.

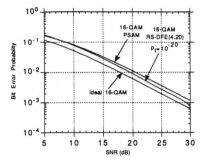

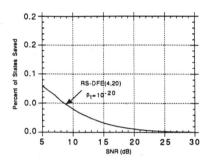

Figure 5. BEP performance of the DFE(4,20) algorithm with 16-QAM modulation. N_p=5 and f_D=.02.

Figure 6 Complexity savings for the RS-DFE(4,20) algorithm with 16-QAM. N_p=5 and f_D=.02.

Recall that the optimum algorithm needs to calculate $p\big(d_\kappa(k)|x(k)\big)$ to form the optimum decision rule. This statistic takes on values for each element of the state space $\Omega_d{}^{K+1}$ and if M or K is large the complexity is prohibitive. Significant complexity reduction can be achieved if the recursion is implemented on a partion of $\Omega_d{}^{K+1}$ having significantly less elements than M^{K+1}. An example of this idea is embodied in reduced state sequence estimation (RSSE) [13]. The problem with RSSE as proposed in [13] is that it trades off decision lag for estimator complexity. To achieve high performance in fading channels a significant decision lag is necessary and so a significant generalization of RSSE is necessary. This is possible in the MAP symbol-by-symbol demodulation with large constellations context by using a multiscale approach. This reduce state estimation technique allows a much greater flexibility in trading off performance versus complexity. The details of this algorithm is in [12].

6. Spatial Diversity Algorithms

Diversity is easily incorporated into the MAP decoding rules. Since the posterior probabilities of the transmitted signals are being computed, any additional information concerning these transmitted signals can easily be incorporated into the decision statistics (the posterior probabilities). Antenna diversity can be viewed simply as expanding the observation vector $X(k)$, and algorithms incorporating this information are extensions of the MAP decoding rules highlighted above. Fig. 7 is a plot showing the BEP with an approximate 16-QAM MAP based demodulator for spatial diversity combining when the antennas are spaced far enough apart to achieve approximate independence. In the case when the antennas are placed closer together

116

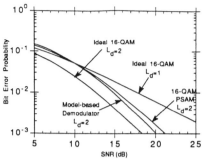

Figure 7 BEP of a MAP based decoder using antenna diversity. 16-QAM. f_DT=0.02.

(i.e, in an antenna array) the spatial coherence of the received wavefront can be exploited. The resulting optimum demodulator effectively forms beams in the directions of arrival of the received signal. This beamforming capability not only improves the demodulation performance but it improves the performance of the link in the presence of cochannel interferers. The characterization of the performance improvement is not complete yet but preliminary results look promising.

7. Decoding with Error Control Coding

Time diversity in mobile communications is typically achieved with interleaving and coding [14]. The use of error control codes adds redundancy in the transmitted signal and interleaving of the transmitted symbols hopefully produces independent fading on each of the code symbols. Dense multiamplitude constellations like 16-QAM have significant practical disadvantages in terms of operation of the power amplifier but these drawbacks can be counteracted with error powerful error control codes [15]. Consequently the model-based demodulation presented above can be significantly generalized if the framework can be extended to incorporate the decoding of error control codes.

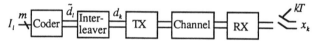

Figure 8. Model for interleaving and coding

To show that this extension to optimum decoding is straightforward consider the block diagram in Fig. 8 of a communication system transmitting m bits per baud with a linear modulation. The optimal information symbol-by-symbol decision rule is given as

$$\hat{I}_l = \arg\max p\big(I_l|\mathbf{x}(\infty)\big) \tag{20}$$

where $\mathbf{x}(\infty)$ represents the observations (i.e., possibly all of the receiver outputs) and I_l is the m bit information symbol transmitted at time l. Defining the following vectors (see (5))

$$\mathbf{I}(k) = \big[I_k, I_{k-1}, ..., I_1\big]^T \qquad \mathbf{I}_n(k) = \big[I_k, I_{k-1}, ..., I_{k-n}\big]^T \tag{21}$$

the optimum K_1 decoding lag decision rule is given as

$$\hat{I}_{l-K_1} = \arg\max \sum_{\mathbf{I}(l-K_1-1), \mathbf{I}_{K_1-1}(l)} p\big(\mathbf{I}(l)|\mathbf{x}(\infty)\big). \tag{22}$$

Again if we use decision feedback this reduces the decision rule to

$$\hat{I}_{l-K_1} = \arg\max \sum_{\mathbf{I}_{K_1-1}(l)} p\big(\mathbf{I}_{K_1}(l)\big|\mathbf{x}(\infty)\big). \tag{23}$$

This can be further reduced to give

$$\hat{I}_{l-K_1} = \arg\max \sum_{\mathbf{I}_{K_1-1}(l)} 1_{\{\tilde{\mathbf{d}}_{K_1}(l)=\mathbf{g}(\mathbf{I}_{K_1}(l))\}} p\big(\mathbf{d}_{K_1}(l)\big|\mathbf{x}(\infty)\big). \tag{24}$$

where 1_Δ is the indicator function and $\mathbf{g}(\bullet)$ represents the mapping generated by the coder. Using the independence of the fading due to interleaving we get

$$
\begin{aligned}
\hat{I}_{l-K_1} &= \arg\max \sum_{\mathbf{I}_{K_1-1}(l)} 1_{\{\tilde{\mathbf{d}}_{K_1}(l)=\mathbf{g}(\mathbf{I}_{K_1}(l))\}} \prod_{i=0}^{K_1} p\big(\tilde{d}_{l-i}\big|\mathbf{x}(\infty)\big) \\
&= \arg\max \sum_{\mathbf{I}_{K_1-1}(l)} 1_{\{\tilde{\mathbf{d}}_{K_1}(l)=\mathbf{g}(\mathbf{I}_{K_1}(l))\}} \prod_{i=0}^{K_1} 1_{\{d_k=\tilde{d}_{-i}\}} p\big(d_k\big|\mathbf{x}(\infty)\big).
\end{aligned}
\tag{25}
$$

Finally asssuming K lag decoding on the channel symbols (i.e., the statistics computed in (6)) gives

$$\hat{I}_{l-K_1} = \arg\max \sum_{\mathbf{I}_{K_1-1}(l)} 1_{\{\tilde{\mathbf{d}}_{K_1}(l)=\mathbf{g}(\mathbf{I}_{K_1}(l))\}} \prod_{i=0}^{K_1} 1_{\{d_k=\tilde{d}_{-i}\}} p\big(d_k\big|\mathbf{x}(k+K)\big). \tag{26}$$

Note that the mass functions needed in (26) are exactly the statistics computed in the optimum MAP symbol-by-symbol estimators of the channel symbols given in Section 3-5.

Equation (26) shows how the MAP symbol-by-symbol demodulation scheme can be generalized to demodulate signals which use interleaving and coding. Note that the result in (26) assumed symbol-by-symbol decoding of information symbols but similar results can be obtained for the more common sequence estimation formulation. The symbol-by-symbol formulation was presented to maintain consistency in the presentation, but sequence estimation is usually more practical to implement. It should also be noted that for interleaving and coding the symbol-by-symbol channel symbol estimation scheme is necessary for a simple decoding architecture.

8. Conclusion

This paper presented a principled approach to narrowband mobile digital communications which provides a unified framework to incorporate diversity combining techniques, array signal processing, forward error control decoding and optimum demodulation in a single demodulation architecture. The complexity of this approach is significantly greater than standard PSAM techniques but the performance gain as well are potentially significant.

9. References

[1] A. Aghamohammadi and H. Meyr, "A New Method for Phase Synchronization and Automatic Gain Control for Linearly Modulated Signals on Frequency Flat Fading Channels," *IEEE Trans. Commun.*, vol. COM-38, January 1991, pp. 25-29.

[2]   M.L. Moher and J.H. Lodge, "TCMP - A Modulation and Coding Scheme for Rician Fading Channels," *IEEE J. Select. Areas Commun.*, vol. SAC-7, December 1989, pp. 1347-1355.

[3]   A. Bateman, "Feedforward Transparent Tone-in-Band: Its Implementations and Applications," *IEEE Trans. on Veh. Technol.*, vol. VT-39, August 1990, pp. 235-243.

[4]   F. Davarian, "Mobile Digital Communication Via Tone Calibration," *IEEE Trans. Veh. Tech.*, vol. VT-36, May 1987, pp. 55-62.

[5]   M. Yokohama, "BPSK System with Sounder to Combat Rayleigh Fading in Mobile Radio," *IEEE Trans. Veh. Tech.*, vol. VT-34, February 1985, pp. 35-40.

[6]   M.P. Fitz, "A Dual-Tone Reference Digital Demodulator for Mobile Digital Communications," *IEEE Trans. Veh. Technol.*, vol. VT-42, May 1993, pp. 156-165.

[7]   K. Abend and B.D. Fritchman, "Statistical Detection for Communication Channels with Intersymbol Interference," *Proc. IEEE*, vol. 58, May 1970, pp. 779-785.

[8]   J.K. Cavers, "An Analysis of Pilot Symbol Assisted Modulation for Rayleigh Faded Channels," *IEEE Trans. Veh. Technol.*, vol. VT-40, November 1991, pp. 686-693.

[9]   R.W. Chang and J.C. Hancock, "On Receiver Structures for Channels Having Memory," *IEEE Trans. Info. Theory*, vol. IT-12, October 1966, pp. 463-468.

[10] P.R. Kumar and P. Varaiya, *Stochastic Systems: Estimation, Identification, and Adaptive Control*, Prentice-Hall, Englewood Cliffs, NJ, 1986.

[11] H.V. Poor, *An Introduction to Signal Detection and Estimation*, Springer-Verlag, New York, 1988.

[12] J.P. Seymour and M.P. Fitz, "Improved Synchronization in the Mobile Communications Environment," Technical Report, School of Electrical Engineering, Purdue University, West Lafayette, IN, 1994.

[13] V.M. Eyuboglu and S.U.H. Qureshi, "Reduced-State Sequence Estimation with Set Partitioning and Decision Feedback," *IEEE Trans. Commun.*, vol. COM-36, January 1988, pp. 13-20.

[14] E. Biglieri, *et al.*, *Introduction to Trellis-Coded Modulations with Applications*, Macmillan, New York, 1991.

[15] P. Ho, J. Cavers, and J. Varaldi, "The Effect of Constellation Density on Trellis Coded Modulation in Fading Channels," *IEEE Trans. Veh. Technol.*, vol. VT-42, August 1993, pp. 318-325.

8

BER Performance of Adaptive RAKE Receiver Using Tap Weights Obtained by POCS Deconvolution Technique

Shehzad Hussain[*]   Zoran Kostić[†]   B. Gopinath[*]

Abstract

The RAKE receiver is used in the multipath channels to make use of the inherent time-diversity effect achieved by the use of a direct sequence spread spectrum (DSSS) communication system. Coherent combining of multipath signals by a RAKE receiver improves system performance. The improvement in performance is based on selecting the stronger paths and rejecting the noisy or weak paths. This criterion requires good estimates of the multipath channel parameters. We obtain multipath parameter estimates by the set theoretic deconvolution technique. The method is based on the constrained iterative deconvolution using the method of Projection Onto Convex Sets (POCS). In this paper we analyze the bit error rate (BER) performance of an adaptive RAKE receiver by computer simulations for a BPSK/DSSS system in an indoor environment. The performance is analyzed for indoor multipath channel models, with/without Inter Symbol Interference (ISI). In order to have a realistic effect, we generate the impulse response of the channels by software package SIRCIM.

For comparison purposes, performance of five receivers was analyzed, i.e., Strongest Path (SP), RAKE with ideal estimates (Ideal), Post Detection Integrator (PDI), RAKE with matched filter estimates (MF), and RAKE with POCS estimates. The simulation results show that the RAKE receiver with tap weights being adjusted by POCS estimates performs close to an ideal RAKE for a sounding signal transmitted at Signal to Noise Ratio (SNR) of greater than 25 dB. It performs about 1-2 dB better than the RAKE with matched filter estimates at BER of 10^{-3}. The performance being superior at low sounding SNR. It was also observed that performance of a PDI receiver was extremely poor, i.e. a simple implementation of RAKE receiver in an indoor multipath environment, having large number of taps and small number of useful paths is not feasible.

1 Introduction

In mobile radio environment the signals arrive at the receiver via various paths. In order to make use of this path diversity, RAKE [1-3] receiver is employed to coherently combine the energies in the multipaths to improve system performance. To resolve multipaths, wideband DSSS technique

---

[*]WINLAB, Dept. of Electrical and Computer Engr., Rutgers University, PO Box 909, Piscataway, NJ 08855-0909

[†]AT&T Bell Laboratories, Holmdel, New Jersey 07733

120

is used. The paths are resolvable if the adjacent paths are separated by at least the spreading sequence bit (chip) duration, T_c. RAKE receiver is implemented as a tapped delay line filter. Optimum performance from RAKE is achieved if the tap weights of the RAKE represent the impulse response of the channel. When impulse response of the channel is time-varying, adaptive scheme is needed to keep track of these variations. Various schemes have been proposed to adaptively track the tap weights, some of these techniques are discussed in [9-12].

In this paper we evaluate the tap weights of the RAKE combiner by set theoretic deconvolution technique, POCS [4-7]. This technique has been extensively used in image processing. POCS application for estimating impulse response parameters of a multipath channel for sonar systems was first studied by Kostić [5]. Its usefulness for DSSS system was studied in [6-7]. A synthetic view of set theoretic estimation, along with an extensive list of references is available in [13].

In section 2, we will discuss the channel/receiver model, section 3 gives a brief description about parameter estimation by POCS. Section 4 discusses the computer simulation and results.

2   Transmitter, Channel and Receiver Model

A block diagram of the simulated system is shown in Fig. 1. Input is the sampled information symbols (±1), multiplied by the spreading sequence, s. Impulse response of the multipath channel is represented by h, n is the additive white Gaussian noise sequence, h_m is the impulse response of the matched filter, matched to the spreading sequence, s. The sampled estimate of the multipath channel is obtained by POCS algorithm and matched filtering are represented by $\hat{h}_{pocs}$ and $\hat{h}_{MF}$ respectively. The processor processes the output of the matched filter depending on five type of receivers discussed latter. The processor output is sent to a detector, which makes a hard decision on the received symbol to be +1 or -1.

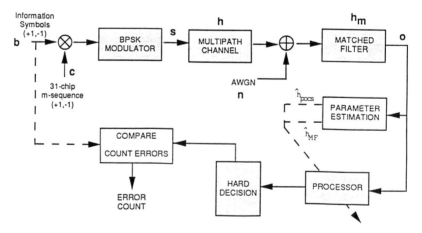

Figure 1: Block diagram of the simulated system

The multipath channel model is expressed as:

$$h(t) = \sum_{n=1}^{N} \alpha_n e^{j\theta_n} \delta(t - \tau_n) \tag{1}$$

where N is the total number of paths, α_n, θ_n and τ_n are the path-strength, phase and excess-delay of the nth path respectively.

In order to perform simulations as close as possible to the real environment, we generate the impulse response of an indoor multipath channel by software package, SIRCIM [8], which in turn is based on extensive indoor empirical measurements. The description of the multipath channel models used in the simulations are as follows.

2.1 Channel Model-1

Channel model-1 simulates a typical office type building with soft partitions. Line of sight (LOS) exists between transmitter and receiver, which are separated by a distance of 25 meters. Average rms delay spread of the channel is 30 ns. Mobile speed is 1 m/s. The impulse response was generated using a software package, SIRCIM [8].

Normalized channel impulse response and its statistics are shown in Fig. 2. Average rms delay spread of the channel corresponds to 3-4 chips duration. Most of the energy in the impulse response is concentrated in the first 12 bins, meaning that there is no ISI. Multipath parameters were estimated at an average SNR of 20 dB and 30 dB. SNR is defined as $20 \log 10(\frac{\|s\|}{\|n\|})$.

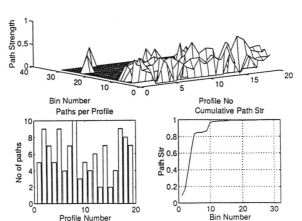

Figure 2: Channel Impulse Response statistics for the simulated channel model-1.

2.2 Channel Model-2

In this model we use the same impulse response as in channel model-1 above but artificially extend it by padding zeros between the paths to simulate ISI. There were two, one, sixteen, and ten zeros padded between first-second, second-third, third-fourth, and fourth-fifth paths respectively. This caused a slight ISI in the two adjacent bits. Multipath channel parameters were estimated by transmitting the sounding signal at an SNR of 20 and 30 dB.

2.3 Channel Model-3

Channel model-3 simulates an open plan building. No LOS exists between transmitter and receiver. Transmitter-Receiver separation is about 22 meters. Average rms delay spread of the channel is 131 ns. Mobile speed is 1 m/s. The impulse response was generated using SIRCIM [8].

Fig. 3 shows the normalized impulse response and its statistics for channel model-3. It is a severe channel, in the sense that it has large number of paths and it is a non LOS channel, in contrast to channel model-1, where 100% LOS exists between transmitter and receiver. Also the energy is distributed over large number of paths causing ISI. Multipath channel parameters were estimated by transmitting the sounding signal at an SNR of 20 and 30 dB.

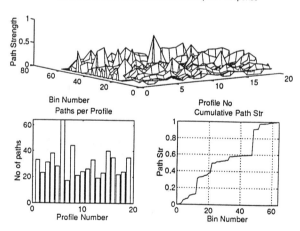

Figure 3: Channel Impulse Response statistics for the simulated channel model-3.

3 Channel Parameter Estimation Using POCS

POCS is an iterative technique to find a feasible solution. A feasible solution is one which satisfies all the constraints placed on the solution. Therefore unlike other signal restoration methods, like maximum likelihood, MAP, or Minimum Mean Squared Error (MMSE) which provide a unique solution, POCS algorithm returns a feasible solution, an element of a family of solutions, satisfying the given constraints.

POCS makes use of the apriori information available about the signal and channel characteristics to define the convex sets of constraints. It was reported in [5] that POCS has the ability to achieve very good resolution in multipath component separation, exact estimation of path delays and accurate estimation of path attenuation factors.

Using matrix representation the output o of the matched filter can be expressed as:

$$o = s \otimes h \otimes h_m + n \otimes h_m \qquad (2)$$

where $\otimes$ represents the convolution operator, o contains the observations at the output of the matched filter.

The observation vector, o, at the output of the matched filter can be rewritten as:

$$o = Ah + v \qquad (3)$$

where $v = h_m \otimes n$, A is the lower triangular convolution matrix or circulant matrix obtained from samples of $s \otimes h_m$.

POCS iteratively estimates h in a linear model described as $o = A h + n$ by sequential projections onto convex sets satisfying the constraints placed on the solution. Brief description about the constraints used and their corresponding projection operators follows:

3.1 Constraints and Projection Operators

3.1.1 Variance of the Residual Constraint, C_v.

The residual, r is given as $r = o - Ah$.

$$C_v = \{h \mid \| o - Ah \|^2 \le \delta_v\} \qquad (4)$$

where δ_v is the bound which can be easily computed from the tables for a required Confidence Interval (C.I.), when the additive noise is assumed Gaussian.

Projection Operator for C_v, P_v

For a given $h \notin C_v$, finding a projected vector $h_p \in C_v$, modify h as:

$$h_p = h + (A^t A + \frac{1}{\lambda}I)^{-1} A^t (o - Ah) \qquad (5)$$

where λ is the solution of $\| o - Ah \|^2 = \delta_v$.

3.1.2 Mean of the Residual Constraint, C_m

$$C_m = \left\{ \mathbf{h} \; \middle| \; \middle| \sum_i (o_i - [Ah]_i) \middle| \le \delta_m \right\}$$ (6)

where o_i and $[Ah]_i$ are the ith elements of the vectors $\mathbf{o}$ and $\mathbf{Ah}$, respectively. The bound δ_m can be obtained from the tables for a required C.I., for the additive Gaussian noise.

Projection Operator for C_m, P_m

$$\mathbf{h}_p = \begin{cases} \mathbf{h} + \left(\frac{\sum r_i - \delta_m}{\|\mathbf{f}_c\|^2}\right) \mathbf{f}_c & \sum r_i > \delta_m \\ \mathbf{h} + \left(\frac{\sum r_i + \delta_m}{\|\mathbf{f}_c\|^2}\right) \mathbf{f}_c & \sum r_i < -\delta_m \\ \mathbf{h} & \text{otherwise} \end{cases}$$ (7)

where $\mathbf{f}_c^t = (\sum_i \mathbf{A}_{i1}, \sum_i \mathbf{A}_{i2}, \dots \sum_i \mathbf{A}_{iN})$

3.1.3 Outliers of the Residual Constraint, C_o

$$C_o = \{\mathbf{h} \mid \mid o_i - [Af]_i \mid \le \delta_o\}$$ (8)

where δ_o is the 3 standard-deviation bound.

Projection Operator for C_o, P_o

$$\mathbf{h}_p = \begin{cases} \mathbf{h} + \left(\frac{r_i - \delta_\rho}{\|\mathbf{f}_i\|^2}\right) \mathbf{f}_i & \text{if } r_i > \delta_o \\ \mathbf{h} + \left(\frac{r_i + \delta_\rho}{\|\mathbf{f}_i\|^2}\right) \mathbf{f}_i & \text{if } r_i < -\delta_o \\ \mathbf{h} & \text{otherwise} \end{cases}$$ (9)

where $\mathbf{f}_i$ is the column vector containing the ith row of the matrix $\mathbf{A}$.

3.1.4 Amplitude Constraint, C_a

$$C_a = \{\mathbf{h} \mid h_i \le \max(\mid \hat{h}_{MF} \mid)\}$$ (10)

Projection Operator for Amplitude Constraint, P_a

$$h_{p_i} = \begin{cases} h_i & \text{if } h_i \le \max(\mid \hat{h}_{MF} \mid) \\ sign(h_i)\max(\mid \hat{h}_{MF} \mid) & \text{otherwise} \end{cases}$$ (11)

3.1.5 Support Constraint, C_s

$$C_s = \{\mathbf{h} \mid h(i) = 0, \; i \notin I\}$$ (12)

where I is the indicator set for the support S, i.e. it specifies the extent (length) of the impulse response sequence.

Projection operator for Support constraint, P_s

$$\mathbf{h}_p = \begin{cases} h_i & \text{if } i \in I \\ 0 & \text{otherwise} \end{cases} \tag{13}$$

First three constraints are based on the additive noise in the system. Fourth constraint is based on the impulse response strength and the fifth constraint is about the maximum delay spread of the channel.

For the given constraints and projection operators the POCS algorithm is expressed as:

$$\hat{\mathbf{h}}_{k+1} = P_o P_m P_v P_a P_s \hat{\mathbf{h}}_k, \qquad k = 0, 1, 2, \ldots \tag{14}$$

where $\hat{\mathbf{h}}_0$ is an arbitrary initialization vector. In these simulations, $\hat{\mathbf{h}}_0$ is selected as the vector o at the output of the matched filter, k denotes the estimate at kth iteration.

4 Simulation and Results

The simulations were performed by intermittently transmitting sounding symbol for each profile followed by 530 information-symbols per profile as shown in Fig. 4.

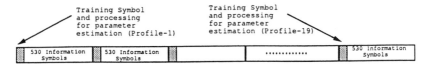

Figure 4: Transmitted signal flow diagram

Using SIRCIM, 19 profiles were generated over a distance of 1 meter for each channel model. Channel impulse response and its statistics are shown in Fig. 2 and Fig. 3. We used this data to measure the performance of five different receivers, discussed later.

To detect each of the 530 transmitted bits per profile, we sample the the output, o, at the ideal sampling instant, i.e., we are assuming perfect synchronization at the receiver. The output sequence, o, is then written in a matrix form, $\mathbf{V}$. Each column of this matrix represents the sampled signal at the output of the matched filter corresponding to each bit. It is on these columns of matrix, $\mathbf{V}$, the five receivers discussed below make their decision regarding the transmitted symbol. The decision mechanism for various receivers is:

4.1 RAKE (Ideal) Receiver

In this receiver, we assume that the impulse response parameters can be accurately estimated. We decide 1 if the $\sum_{i=1}^{lv} \mathbf{h}^*(i)\mathbf{V}_k(i)$ is positive, otherwise the receiver decides that a 0 is received. where lv is the length of the vector $\mathbf{V}_k$, k represents the kth column vector of matrix $\mathbf{V}$ and i represents the ith element of vector $\mathbf{V_k}$, h˙ is complex conjugate of the channel impulse response. Fig. 5 shows the block diagram of the RAKE receiver.

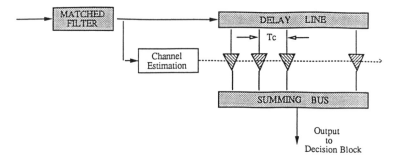

Figure 5: RAKE Receiver

4.2 Post Detection Integrator (PDI) Receiver

PDI receiver makes a decision after integrating over the full multipath spread or for the discrete case, summing all the elements of vector $\mathbf{V_k}$. If $\sum_{i=1}^{lv} Re(\mathbf{V}_k(i)e^{-j\theta_i})$ is positive, 1 is decided, else 0. PDI receiver has the same configuration as a RAKE receiver, except that all taps have unity gain (see Fig. 5).

4.3 Strongest Path (SP) Receiver

In this receiver the decision about, whether 1 or 0 is received is made on the result of $\hat{\alpha}_i e^{-j\theta_i}\mathbf{V}_k(i)$. Decision of 1 or 0 is made on whether the result is positive or negative, i represents the location of the maximum absolute value of the impulse response estimate of the channel at the output of the matched filter.

4.4 RAKE (POCS estimates) Receiver

In this case the impulse response parameters are estimated using POCS deconvolution technique. The decision of 1 is made if $\sum_{i=1}^{lv} \hat{h}_{POCS}^*(i)\mathbf{V}_k(i)$ is positive, otherwise the receiver decides 0. The complex conjugate estimated impulse response of the channel, $\hat{h}_{POCS}^*$, is obtained by POCS deconvolution algorithm.

4.5 RAKE (Matched Filter estimates) Receiver

In this case the output of the matched filter is taken as the estimate of the impulse response parameters and the receiver decides on a 1 or 0 depending on the sum, $\sum_{i=1}^{lv} \hat{h}_{MF}^*(i)\mathbf{V}_k(i)$, is positive or negative respectively, $\hat{h}_{MF}^*$ is the complex conjugate estimate of the channel impulse response at the output of the matched filter.

127

4.6 Results

Fig. 6 shows the BER curves for the five receivers discussed above for **channel model-1**. There was no ISI and POCS receiver performed about 1 dB better than MF receiver at BER of 10^{-3}, when multipath parameters were estimated at SNR of 20 and 30 dB. POCS receiver performed equally well as an ideal receiver, when multipath parameters were estimated at SNR of 30 dB. The performance of POCS receiver in comparison to non-RAKE (Strongest Path) receiver is more than 5 dB superior at BER of 10^{-3}. The performance of PDI receiver is extremely poor in such an environment.

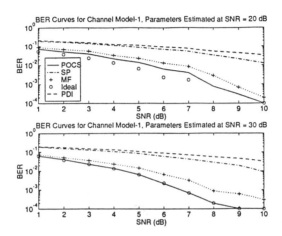

Figure 6: Channel Model-1. BER curves for 5 simulated receivers. (**Legend** solid: POCS, o: Ideal, +: Matched Filter, dash: PDI, dash-dot: Strongest Path)

Fig. 7 shows the BER curves for the five receivers discussed above for **channel model-2**. Note that in this case, ISI was intentionally introduced. Performance of all receivers was slightly degraded, compared to performance of channel model-1. POCS receiver performance is close to an ideal receiver for parameter estimation at SNR of 30 dB. POCS performance in comparison to MF receiver is better by 0.5 to 1.5 dB at BER of 10^{-3}. PDI performance was the worst as is true in channel model-1 discussed above.

Fig. 8 shows the BER curves for the five receivers discussed above for **channel model-3**. It was a non LOS channel with energy distributed over many paths causing ISI. The MF and POCS receivers performed equally well, when multipath parameters were estimated at SNR of 20 dB. The performance was approximately 2 dB worse than an ideal receiver. However, for parameter estimation at SNR of 30 dB, both MF and POCS performed equally well as the ideal receivers would, meaning that for such channels any processing done to obtain better estimates of multipath channel parameters is not very useful. The performance of PDI and non-RAKE (SP) was poor.

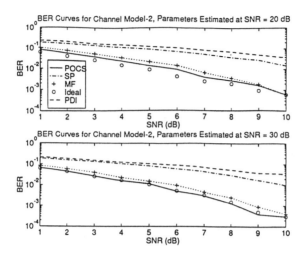

Figure 7: Channel Model-2. BER curves for 5 simulated receivers. (**Legend** solid: POCS, o: Ideal, +: Matched Filter, dash: PDI, dash-dot: Strongest Path)

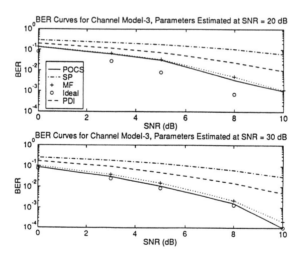

Figure 8: Channel Model-3. BER curves for 5 simulated receivers. (**Legend** solid: POCS, o: Ideal, +: Matched Filter, dash: PDI, dash-dot: Strongest Path)

5 Conclusion

In this paper, computer simulation results have been reported for BPSK/DSSS system, showing bit error rate performance of five receivers in an indoor multipath propagation environment. The main purpose was to see as to how the impulse response parameters estimated by the POCS deconvolution algorithm compare with error rate performance of a conventional RAKE receiver using ideal estimates or matched filtering estimates. The simulation showed us that POCS receiver can provide a gain of 0.5-1.5 dB over the conventional RAKE receiver at BER of 10^{-3}, when multipath channels are estimated at SNR of 20 and 30 dB. POCS receiver performance is close to an ideal RAKE receiver when multipath channel parameters are estimated at SNR of greater than 25 dB.

Simulation results show that considerable gain in BER performance is achievable by a RAKE receiver in multipath environment, if multipath parameters can be accurately estimated at low SNR (below 15 dB). However MF receiver obtains performance close to an ideal RAKE receiver, when sounding signal is transmitted at SNR of greater than 30 dB, thus suggesting that any additional processing done to achieve better multipath parameter estimates is not very useful.

Simulation results also show that for multipath channels with large delay spread and small number of useful paths, a simple configuration of RAKE receiver (PDI), performs even worse than a non-RAKE receiver. Under such channel conditions, some form of threshold is required to eliminate contribution of noisy taps, to achieve a reasonable performance from a simple RAKE (PDI) configuration.

References

1. R. Price et al., *A communication technique for multipath channels*, Proc. IRE, March 1958, pp. 555-570.

2. J. G. Proakis, *Digital Communications*, 2nd ed., McGraw-Hill, New York, 1989.

3. G. L. Turin, *Introduction to spread-spectrum antimultipath techniques and their application to urban digital radio*, Proc. IEEE, vol. 68, no. 3, March 1980, pp. 328-353.

4. H. J. Trussell et al., *The feasible solution in signal restoration*, IEEE Trans. ASSP, vol. 32, no. 2, April 1984.

5. Z. Kostić et al., *Estimation of parameters of a multipath channel using set-theoretic deconvolution*, IEEE Trans. Commun., vol. 40, no. 6, June 1992.

6. Z. Kostić et al., *Resolving sub-chip spaced multipath components in CDMA communication systems*, Proc. IEEE VTC, 1993, pp. 428-431.

7. S. Hussain, *Estimating complex impulse response of a multipath channel using POCS deconvolution*, WINLAB Internal Report, Rutgers, The State University, Oct. 1992 and Jan. 1993.

8. T. S. Rappaport et al., *SIRCIM: simulation of indoor radio impulse response models*, 1990, VTIP Inc.

9. C. N. Pateros et al., *Adaptive correlator receiver performance in fading multipath channels*, Proc. IEEE VTC, 1993, pp. 746-749.

10. Y. Sanada et al., *Adaptive RAKE system for mobile communications*, IEEE Intl. Conf. on Selected Topics in Wireless Communications, Vancouver, Canada, June 1992, pp. 227-230.

11. P. M. Grant et al., *Performance of a spread spectrum receiver design*, IEEE 2nd Intl. Symp. on Spread Spectrum Techniques and Applications, Yokahoma, Japan, Nov. 1992.

12. P. A. Bello, *Performance of some RAKE modems over the non-disturbed wide band HF channel*, IEEE Mil. Commun. Conf., San Diego, California, Oct. 1988, pp. 89-95.

13. P. L. Combettes, *The foundations of set theoretic estimation*, Proc. IEEE, vol. 81, no. 2, Feb. 1993, pp. 182-208.

9

A Macrocell Model Based on the Parabolic Diffusion Differential Equation

Jan-Erik Berg

Ericsson Radio Systems AB
S-164 80 Stockholm
Sweden

Abstract

A method to determine the path loss in macrocells, based on the parabolic diffusion differential equation, disregarding the phase information of the propagating wave, is suggested. The multiple knife-edge approach is applied and only multiple diffracted waves, no reflections, are considered. A non-flat terrain can be handled. The differential equation is solved by using the explicit Forward-Difference method, where the distance between the calculation points can be 5 m in the height direction when the wavelength is only 0.3 metres, which makes the method extremely computer efficient.

1   Introduction

In some papers, the path loss in macrocells is described as a multiple diffraction phenomenon, where the buildings are replaced by absorbing half-screens [1-2,8]. The same approach is used for the model described in this paper. However, the mathematical method is totally different and is not based on the wave equation. The model was first presented in [3] and some extended examples were given in [4], where the method was entirely based on a heuristic approach. The method has now been slightly modified and extended to handle quite arbitrary terrain profiles.

It will be shown that with some simple assumptions, the use of the diffusion or heat differential equation can, to some extent, be theoretically supported. An explicit expression, which is a very accurate approximation of the solution of the suggested differential equation for the single knife edge case will then be given. The expression is reciprocal and has the same general average behaviour close to and above the transition zone, between line of sight (LOS) and none line of sight (NLOS), as the well known scalar solution of the diffracted field after a totally absorbing half-screen. Below the transition zone, an entirely empirical approach is applied.

For the practical multiple screen case, the differential equation is numerically solved by using the simple explicit Forward-Difference method. The method will be compared with the well-known empirical Hata model and with measurements at 970 MHz along a 10 km hilly terrain profile.

2   General Model Approach

An example of a terrain profile with some buildings and trees is shown in Fig.1. The profile is replaced by absorbing half-screens, where the edges of the screens coincide with the average height of the surrounding buildings or trees. An important point is that the chosen location of each screen does not have to coincide with an existing building or tree. The distance between the screens is a parameter with which the path loss can be changed. Different kind of environments can then be modelled with different screen separations, which can vary along the terrain profile.

A wave is assumed to propagate from the base station (Bs) antenna. When the propagating wave hits the first knife-edge, location A in Fig.1, its amplitude is set to zero below the edge due to the assumed absorb-

132

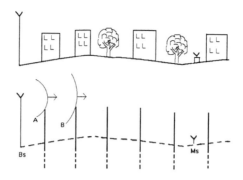

Fig. 1. Original and simplified terrain profile.

ing properties of the screen. The wave is then propagating towards location B, obeying the suggested diffraction law, where the same procedure is repeated. Thus with this repeated procedure, the wave will propagate along the total length of the terrain profile, considering the height variation along the total path. For any mobile station location (Ms), the path loss between the antennas can be determined during this process.

3   The Differential Equation

Assume that a line source is located at position O, see Fig. 2, producing a flow of fluid with the density $\rho = \rho(r, \theta)$ with the velocities $v_{\hat{r}}$ and $v_{\hat{\theta}}$, and that the flow in the θ direction is according to the following law

$$\rho \cdot v_{\hat{\theta}} = -C_1 \cdot \frac{\partial \rho}{\partial s} \qquad (3.1)$$

where $s = r \cdot \theta$. Further, assume that the velocity in the r direction is constant

$$v_{\hat{r}} = C_2 \qquad (3.2)$$

Now, by applying the well known equation of continuity and assuming a steady flow, $\partial \rho / \partial t = 0$, the following equation, in cylindrical coordinates, will be obtained

$$\frac{\rho}{r} + \frac{\partial \rho}{\partial r} = C \cdot \frac{1}{r^2} \cdot \frac{\partial}{\partial \theta}\frac{\partial \rho}{\partial \theta} \qquad (3.3)$$

where $C = C_1 / C_2$.

If $u(r, \theta)$ is introduced with the relationship $\rho = u/r$, then equation (3.3) becomes (with simplified derivative notation)

$$u_r = C \cdot \frac{1}{r} \cdot u_{\theta\theta} \qquad (3.4)$$

which can be written

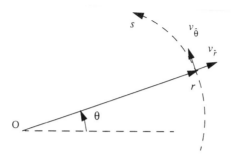

Fig. 2.   Definition of coordinate system.

$$u_r = C \cdot u_{ss} \tag{3.5}$$

where $u = u(r, s)$. Expression (3.5) is identical with the parabolic diffusion, or heat, differential equation if the variable r is interpreted as the time t.

Now, the solution $\rho(r, \theta)$, which determines the density of the fluid originating from a line source, is interpreted to be proportional to the power density for a cylindrical radio wave from a line source. Finally, it is assumed that the power density from a corresponding point source is proportional to $\rho(r, \theta)^2$

The different steps above, describe in a mathematical form the kind of analogy which has been used in order to suggest the use of (3.5) to solve wave propagation phenomena in macrocellular environments. The solution which satisfies (3.5) has some desired properties, which is the main reason why it has been chosen. The most important point is of course that the method works and is simple to use, then it is unessential that the interpretation is contradictory to Maxwell's equations.

4   The Single Knife Edge Case

For the case with one absorbing screen or knife-edge, see Fig. 3, a very accurate solution of (3.4) for $r \geq R_0$, which fulfills the boundary condition $u(r = R_0, \theta < 0) = 0$ and $u(r = R_0, \theta > 0) = 1$, is

$$u(r, \theta) = cnorm\left(\frac{r \cdot \sin(\theta) \cdot R_0 \cdot \sqrt{R_0 + R_1}}{(R_0 + R_1) \cdot \sqrt{2 \cdot C \cdot (r - R_0) \cdot (r - R_1)}} \right) \tag{4.1}$$

where $cnorm$ is the standard normal distribution function. The equation is reciprocal because $R_0 \cdot \sin(\theta) = R_1 \cdot \sin(\alpha)$.

The solution for the diffracted field of a scalar wave after an absorbing screen is given in every standard book on this topic, see e.g. [5]. The parameter v is often used when describing the diffracted loss relative free space loss

$$v = \pm\sqrt{R_0 + R_1 - r} \cdot \frac{2}{\sqrt{\lambda}} \tag{4.2}$$

where λ is the wavelength and the minus sign is valid in the NLOS region, $\theta < 0$.

It is quite easy to show that the argument of the solution in (4.1) is approximately proportional to the parameter (4.2). This agreement is perfect when $|\theta|$ becomes small. It is straightforward to show that the

134

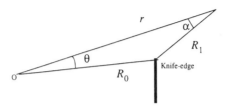

Fig. 3. Distances and angle definition for the single knife-edge case.

derivative with respect to θ of the correct solution at $\theta = 0$ is identical with the corresponding derivative of (4.1) when

$$C = \frac{\lambda}{2 \cdot \pi} \tag{4.3}$$

Thus, this value of C has been chosen and has been used for all of the results described below. It should be noted that this value of C might not be the optimum value.

With $v = v(r, \theta)$ and using (4.3), it can be shown that (4.1) can be approximately written

$$u(v(r, \theta)) \approx cnorm\left(v(r, \theta) \cdot \sqrt{\frac{\pi}{2}} \right) \tag{4.4}$$

The correct solution of the diffracted scalar field and the result according to (4.4) is given in Fig. 4. As can be seen, they have the same general average behaviour. However, the deviation is unsatisfactorily when $v \ll 0$.

5   The None Line Of Sight Region

To solve the problem with the deviation in the NLOS region, described above, an entirely empirical approach has been used. To start with, an approximate solution of the diffracted scalar field in the NLOS region is applied

$$u_{NLOS}\left(r, \frac{s}{r} \right) = \frac{1}{2} + atan\left(v\left(r, \frac{s}{r} \right) \cdot \frac{\pi}{2} \right) \cdot \frac{1}{\pi} \tag{5.1}$$

Secondly, at the transition border between LOS and NLOS, the derivative, u'_0, with respect to s, and the actual level, u_0, is determined at the distance r. For the simple case with one knife edge, the derivative and the actual level can be determined directly from (4.1) or (4.4) at $s=0$. For the multiple knife edge case, the transition border is defined by the edge from the last half screen. The derivative is then numerically determined from the calculated results on and above this border, not below. The values below the actual border is then determined by using the following equation

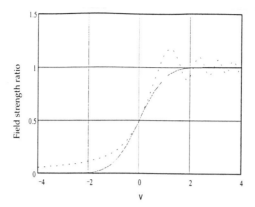

Fig. 4. Field strength ratio of diffracted field after a single knife-edge as a function of the diffraction parameter v. Correct solution, dashed line, and result according to (4.4), solid line.

$$\tilde{u}_{NLOS}(r, s) = u_{NLOS}\left(\Gamma, \frac{s}{\Gamma}\right) \cdot u_0 \cdot 2 \tag{5.2}$$

where

$$\Gamma = \frac{R_0}{2} \cdot \left(1 + \sqrt{1 + \frac{8 \cdot u_0^2}{\lambda \cdot R_0 \cdot u'_0^2}}\right) \tag{5.3}$$

With the suggested method, the derivative with respect to s will be continuous at the border. For the single knife edge case and below the border, the result will be within about 1 dB compared to the correct solution. Further, for the single knife edge, the second derivative with respect to s will be equal to zero at the border, which is the case for the correct solution.

6 The General Multiple Knife Edge Case

To start, at discrete locations along a limited part of the circle which passes through the first knife edge, the values of u are set to one above and on the edge, and zero below the edge. These locations must be chosen with constant distance Δs, see Fig.5. One of these locations must coincide with the knife edge.

Now, assume that the values have been determined along the curve with the radius r_j at locations with constant distance Δs, see Fig. 5. The distance to the next knife-edge is r_{j+1} and to determine the levels along the arc at this distance, equation (3.5) is used. To solve this numerically, the well known explicit Forward-Difference method is applied. Due to this discrete method, the value corresponding to the location on the edge at the distance r_j must first be multiplied with $1/2$. The values along the new arc, r_{j+1}, is then determined with the following formula

$$u_{n, j+1} = \beta \cdot (u_{n+1, j} - 2 \cdot u_{n, j} + u_{n-1, j}) + u_{n, j} \tag{6.1}$$

where

136

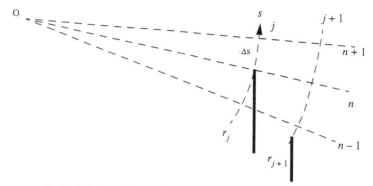

Fig. 5. Definition of the coordinate system for the multiple knife-edge case.

$$\beta = C \cdot \frac{(r_{j+1} - r_j)}{(\Delta s)^2} < \frac{1}{2} \tag{6.2}$$

In order to get a stable solution, the relational condition in (6.2) must be fulfilled.

The values along the curve at the distance r_{j+1} have now been determined. The derivative, u'_0, with respect to s at the distance r_{j+1} is then numerically determined at the position with the same n value as the position which at the distance r_j was located on the edge. For the example shown in Fig. 5, the derivative shall be determined at position $(n, j+1)$. The derivative must be calculated by using the discrete values on and above, not below, this location. Now, below this location new values are determined by using (5.2) and (5.3) where R_0 is replaced by r_j. Above this location, an interpolation method must be applied in order to determine the new values. These new values must be calculated at locations with constant distances Δs. As mentioned above, one of these new positions must be located on the new edge. The values at the positions below the new edge must of course be set to zero.

Now it is only to repeat the same procedure in order to determine the values along the next curve. Thus, the model is entirely based on (6.1) and (5.2), which are very simple to use.

A typical value of Δs at 900 MHz is 5 metres, and the maximum length of the arc should be around 100 metres if the terrain profile is 10.000 metres. This length must of course be less at shorter distances and it can then grow when the distance increases. The screen separations should be in the range from about 100 to 200 metres. These figures and the simplicity of the equations (6.1) and (5.2) makes the method extremely computer efficient.

From the calculated value $u_{j,n}$ at the distance r and location s, the total path loss between isotropic antennas is then determined according to

$$L_{dB} = 20 \cdot \log\left(\frac{4 \cdot \pi \cdot r}{\lambda \cdot u_{j,n}}\right) \tag{6.3}$$

It should be noted that according to the method described above, the wave is always assumed to propagate in the direction from the source. This is of course unrealistic far below a knife edge, but in practice this is not a problem for a majority of common terrain profiles. However for very hilly terrain with steep slopes, the suggested method must be modified, else the calculated path loss will become too pessimistic.

7   Model Parameters

The main parameters of the model is the screen distance w and the distance w' between the mobile antenna

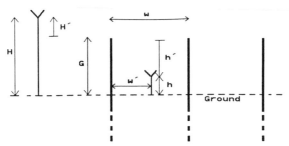

Fig. 6. Definition of model parameters.

and the closest screen towards the base station antenna, see Fig. 6. and Fig. 1. If the screen distance w decreases, then the resulting path loss increases. The same condition is of course valid for the parameter w'. This latter parameter does also determine the path loss dependency as a function of the mobile antenna height

The actual base antenna height H and mobile antenna height h should be used. However, one must be aware of that the resulting path loss is more strongly correlated to the relative antenna heights H' and h', than to H and h.

Further, the height of the knife edges, G, should describe the average height of the obstacles, e.g. buildings or trees, surrounding the actual screen location. As mentioned earlier, there is no need to locate the screens at the same position as the existing buildings. However, an exclusion from this rule should be considered if the base station coverage distance is less than up to about 200 to 300 metres, because in that case the location of every single building becomes important and a statistical approach might not be appropriate.

Another parameter, which is not shown in Fig. 6, is the distance Δs between the calculation points along the arc. In all the examples given below, Δs=5 metres has been used.

At last, the constant C in (3.3-3.5) can be considered as a parameter, however in all of the examples below the value of C has been chosen according to (4.3).

8   Comparison with Empirical Model

The path loss according to the Hata model [6] in urban areas at 900 MHz is shown in Fig. 7 for 30 and 60 metres base antenna height and 2 metres mobile antenna height. Also, in the same figure, is given the path loss according to the suggested model for 15, 30 and 60 metres base antenna heights. The Hata model is not valid for base antenna heights lower than 30 metres and it is not applicable when the base antenna is close to the height of the surrounding buildings.

For the suggested model it was assumed that the height of the buildings, G, was 15 metres. The parameter values w=200 and w'=25 metres were chosen. Further, the curvature of the earth was considered, and an earth radius factor of 1.33 was used.

The calculated path loss as a function of the mobile antenna height h at a distance of 5000 metres is illustrated in Fig. 8 for the case with 30 metres base antenna height. As can be seen in the figure, the suggested model gives a continuous result at the line of sight border, $h = G = 15$ metres. The mobile antenna height dependency according to the Hata model is also shown in the same figure. The agreement is quite good in the range where the Hata model is valid, $h < 10$ metres.

These examples show that by choosing quite realistic parameter values, the resulting path loss variation will be quite close to the expected levels. It should be noted that the results given in Fig. 7 and Fig. 8 are not based on an optimal adaptation of the suggested model to the Hata model.

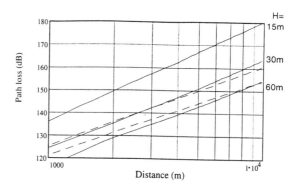

Fig. 7. Path loss according to the suggested model, solid lines, and the Hata model, dashed lines, in urban areas at 900 MHz as a function of the distance for different base antenna heights, H.

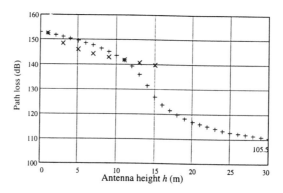

Fig. 8. Path loss according to the suggested model (+) and the Hata model (x) in urban areas at 900 MHz and 5000 metres from the base station as a function of the mobile antenna height h. Free space path loss is shown as reference, dashed line.

9   Comparison with Measurements

A comparison between the suggested model and measured results at 970 MHz from an open hilly terrain in Denmark [7] is shown in Fig. 9. The measurements were performed along an about 10.000 metres long almost straight road with a very low base antenna height of 10.4 metres, which is a difficult case to predict, and with a mobile antenna height of 2. 4 metres. The terrain height variation is displayed in Fig. 9 together with measured and predicted path loss.

A screen separation of $w = 200$ metres was chosen and $w' = 25$ metres was used in the model. Due to that the road was not perfectly straight, a knife edge height of $G = 1$ metre was used in order to consider small obstacles close to the road. The actual antenna heights were applied and an earth radius factor of 1.33 was used.

Considering the simplicity of the model, the very low base antenna height and the fact that reflections

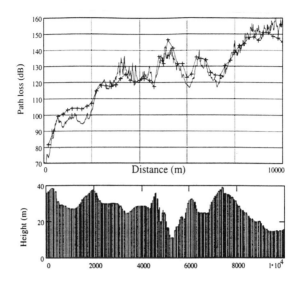

Fig. 9.   Below, terrain height profile. Above, path loss according to the model (+) and measured path loss at 970 MHz along the terrain profile.

are not considered in the model, it is remarkable how well the computed losses follow the measured results. The distance resolution of the measurements was 10 metres while the loss was computed every 200 metres. By interpolation, a predicted value was determined every 10 metres, the resulting standard deviation of the difference between the measured and predicted values was found to be 5.4 dB.

The time consumption to predict the losses along the 10.000 metres terrain profile took less than one second on a 486 PC (66 MHz).

10   Conclusion

It has been demonstrated that

*   with some simple assumptions, the use of the diffusion equation, to determine the path loss in macrocells, and disregarding the phase information of the propagating wave, can to some extent be theoretically supported.

*   the approach to treat the wave propagation along a terrain profile as a result of multiple diffraction from absorbing half-screens and disregarding reflections, seems to be a successful method.

*   the method can be applied in open areas, where the physical obstacles are quite small.

*   the method is extremely computer efficient.

*   in comparison with the empirical Hata model at 900 MHz, the agreement was quite good regarding absolute level, distance, base antenna height and mobile antenna height dependence, when reasonable parameter values were applied.

*   the agreement between measurements and the model along a 10.000 metres road in open hilly terrain at 970 MHz with a low base antenna height of 10.4 metres was also good and with a standard deviation of the error of 5.4 dB.

Acknowledgements

The author wish to thank Prof. J.B. Andersen and his group at the University of Aalborg, Denmark, for providing the measurement data and the corresponding terrain profile.

References

[1] J. Walfisch and H. Bertoni, "A Theoretical Model of UHF Propagation in Urban Environments," *IEEE Transactions on Antennas and Propagation,* Vol. 36, No 12, pp.1788-1796, Dec. 1988.

[2] "A Diffraction Based Theoretical Model for Prediction UHF Path Loss," *IEEE Trans. Veh. Technol.,* Vol. 37, pp. 63-67, 1988.

[3] Jan-Erik Berg, "A Macrocell Model Based on the Parabolic Heat Differential Equation", *COST 231 TD(92) 6,* Vienna, Jan. 1992.

[4] Jan-Erik Berg "A Macrocell Model Based on the Parabolic Heat Differential Equation," *Proceedings of Nordic Radio Symposium,* Aalborg Denmark, pp. 39-42, June 1-4, 1992.

[5] "Transmission Loss Predictions for Tropospheric Communication Circuits", *National Bureau of Standards, Technical Note 101,* Vol. 1, May 7, 1965.

[6] M. Hata, "Empirical Formula for Propagation Loss in Land Mobile Radio Services, "*IEEE Trans. Veh. Technol.,* Vol. VT-29, No. 3, pp. 317-325, Aug. 1980.

[7] J.B Andersen, J.T. Hviid and J. Toftgård, "Comparison Between Different Path Loss Models," *COST 231 TD(93)-06,* Barcelona, Jan. 1993.

[8] Jan-Erik Berg and Håkan Holmquist, "An FFT Multiple Half-Screen Diffraction model", to be published in *Proceedings VTC 94,* Stockholm, Sweden, June 7-10, 1994.

10

A Nano-Cellular Local Area Network Using Near-Field RF Coupling

Alan Demers, Scott Elrod, Christopher Kantarjiev
Edward Richley
Xerox Palo Alto Research Center
3333 Coyote Hill Rd.
Palo Alto, CA 94304

March 31, 1994

Abstract

This paper describes a new type of wireless LAN based on near-field RF coupling. The system exploits the rapid spatial decay of field strength within the near-field to provide high isolation between adjacent cells. We believe that this Nano-Cellular system offers unique advantages for mobile computing in an indoor environment. At the operating frequency of 5.3 MHz, the cell size is approximately that of a single office. Small cells offer several benefits, including the possibility of channel re-use and the ability to provide information about user location. Despite the low operating frequency, the available bit rate of 250kbps is sufficient for many practical applications. The radio wavelength of 60m is sufficiently long to eliminate standing wave problems which often plague indoor UHF and microwave systems. The transceiver uses very little power, and has a simple, low-cost design.

The operational theory of the near-field radio is presented, together with detailed measurements of signal strength and packet reception in an indoor environment. The choice of operating frequency is explained as a tradeoff among cell size, bandwidth and FCC limitations. Details of the radio hardware are given, including modulation techniques and antenna optimization. The characteristics of indoor RF noise sources are described, along with the implications for the placement of base stations. Measured results are presented for several geometries where packet loss can result from contention among devices.

Efficient, fair channel allocation among multiple clients is provided by a new Media Access Control protocol that takes advantage of the high adjacent-cell isolation provided by near-field coupling. This protocol is described, and detailed measurements are presented for several configurations.

1 Introduction

In recent years, various wireless Local Area Networks have been proposed or implemented[1][2]. Major considerations in the design of such networks are bit rate, range, infrastructure complexity, cost, and battery life. This paper compares the design features of existing wireless LANs with a new technique based on near-field RF coupling. We believe that this Nano-Cellular system offers unique advantages for mobile computing in an indoor environment.

For system simplicity, it is desirable to have a minimum set of communication channels. If the range of each cell is limited, channel re-use is possible. The simplest system would have one common channel, but would require a high degree of isolation between cells. Wireless infrared systems [2] [3] have the desirable feature of limited range (opaque walls form very convenient cell boundaries), but they tend to require line-of-sight conditions. For this reason, an approach based on radio frequency transceivers has been pursued.

Experiments [4][5][6] have shown that the UHF and microwave bands allocated for mobile communications [7] have propagation characteristics which make them inappropriate for use inside a building. For several reasons, it is very difficult to obtain the desired cell definition. In part, this is due to the gradual spatial decay of field strength ($\frac{1}{r}$). This slow decay limits the packing density of cells on a common channel. Furthermore, metal objects, such as office furniture and wall framing, cause reflections of the propagating radio waves, leading to distortions in the cell shapes. Finally, the reflections result in standing wave patterns with quite large peak-to-null ratios (typically greater than 20db) over distances of only a few centimeters. As a result, the signal strength varies with position in a highly non-monotonic manner as one recedes from the center of a cell. In practice, this means that for base stations operating on the same channel, very slight movements of a mobile receiver can cause the communication channel to jump between cells in an unpredictable manner.

In order to combine the advantages of small, well-defined cells and to overcome the line-of-sight limitations of infrared, we have developed a new type of cellular LAN based on near-field RF coupling. Near field coupling exhibits a more rapid spatial decay of field strength ($\frac{1}{r^2}$ or $\frac{1}{r^3}$), thereby permitting close packing of cells, all of which operate on a common carrier frequency. The communication channel is identical for all cells and is non frequency agile. At the low carrier frequency of 5.3 MHz, the wavelength is long enough (60 m) that standing wave problems are eliminated. Adequate channel bit rate is available (250kbps) for many practical applications. The RF hardware has low cost and consumes very little power. Protocols have been developed for efficient channel allocation among many clients.

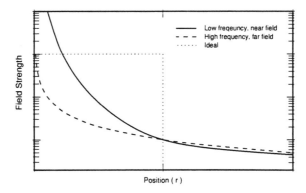

Figure 1: A theoretical comparison among field strength patterns for the ideal, far-field, and near-field systems. Standing wave effects have been ignored for clarity.

2   The Physics of Near-Field Coupling

In contrast with the typical goals of RF communication links, the Nano-Cellular system requires transceivers which have short range and well-defined coverage areas. Isolation between cells is

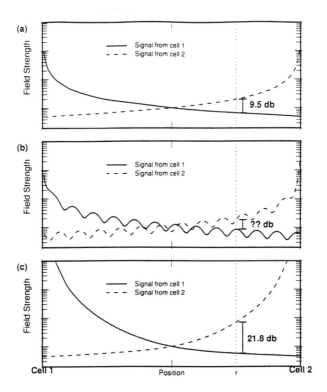

Figure 2: Adjacent cell intereference for far-field and near-field coupling methods. At position r the relative signal strength of the two cells is shown. At (a) is shown the free space far-field case. At (b) is the more realistic case with standing wave effects. The relative signal strength in this case is uncertain as it varies greatly, and non-monotonically, with small changes in position. At (c) is shown the near-field case.

necessary for efficient channel re-use. Ideally, the signal strength of a transmitter would be uniform within the cell and drop to zero outside. While such a pattern is physically impossible, a compromise can be made which will improve isolation between cells over that obtainable with conventional methods.

In a conventional RF system, the receiving antenna is located in the far-field region of the transmitting antenna. This is generally desirable when range is to be maximized. For UHF and microwave transmitters, the far-field region is well established beyond several centimeters from the transmitting antenna. In this region, field strength decreases with distance as $\frac{1}{r}$, where r is the distance from the transmitting antenna. However, all antennas exhibit a near field region as well. Within this region, the field strength follows a much more precipitous rate of decay. Exact definition of the near field region will vary with antenna geometry, but for most purposes, it is approximately defined by $0 < r < \frac{\lambda}{2\pi}$ where λ is the free space wavelength of the signal. Figure 1 shows a comparison among ideal, far-field, and near-field patterns.

The rapid spatial decay of field strength in the near-field region provides a mechanism by which the isolation between cells can be improved over that of a far-field system. Figure 2 shows this improvement by comparison of adjacent-cell interference in near-field and far-field systems. Also

144

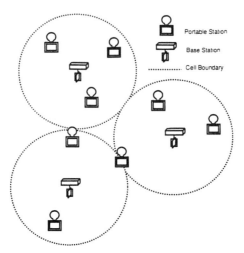

Figure 3: Schematic representation of a Nano-Cellular network showing multiple cells, each with a base station, and multiple portable stations.

shown in Figure 2 is the non-monotonicity exhibited by a UHF far-field system. The standing wave patterns produced at these high-frequencies make adjacent cell interference very irregular over distances much less than the width of a cell. In contrast, near-field operation is obtained by utilization of frequencies significanly lower than those in the UHF. Thus, standing wave patterns are absent in plot (c) of Figure 2.

In the near-field region, the phase relation and relative orientation between electric and magnetic fields is not constant with position. As a result, the Poynting vector has an imaginary component. Energy transfer in the near-field region is accomplished by electric or magnetic coupling between source and sink, rather than by propagation from a transmitter to a receiver. For example, a loop antenna connected to a receiver in the near-field region will extract energy from the magnetic field components, and is hence coupled to the transmitter in much the same way that a transformer

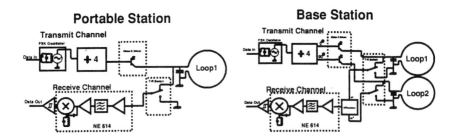

Figure 4: Block diagrams of near-field transceivers for portable stations and base stations. Aside from the quadrature circuitry in the base station transceiver, the circuits are identical.

Top view:

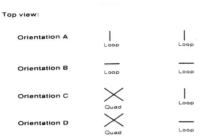

Figure 5: The basic orientations used for measurements with loop and quadrature antennas, as seen in a top view.

secondary is coupled to its primary.

In order to utilize the near-field region, one must choose a signal wavelentgh (λ) appropriate for the desired cell size. For office-sized cells, this constrains the carrier frequency to be roughly 5 to 10 Mhz. A value of 5.3Mhz was chosen for this work. Although other carriers may be feasible, it is not likely that they will differ significantly from this value. Thus, a near-field system with office-sized cells will operate in the HF band, and not in the VHF or UHF bands. A further advantage of operation in this band is that standing wave effects are virtually eliminated.

An obvious disadvantage of such a low carrier frequency is the limitation placed on system bandwidth. The dispersion introduced by the coupling characteristics outlined in the above equations prohibits the channel bandwidth from exceeding about 10% of the carrier frequency. The bit rate associated with this bandwidth depends on the modulation scheme employed. With narrow band FSK having a deviation of 200kHz, a bandwidth of 0.5 Mhz has been found to be adequate for a bit rate of 250kbps.

Although the size of the near-field varies inversely with frequency, there are other considerations. In particular, FCC part 15.223 [7] places limits on the *far-field* signal strength from intentional radiators in such a manner as to allow stronger near-fields as carrier frequency is reduced. This regulatory consideration restricts the dipole strength, IA, which can be used for each transmitter. For example, at 5.3Mhz, IA is limited as follows:

$$IA \leq .0081 \tag{1}$$

where IA has units of (A-m^2). Thus, area and current can be traded according to the above rule.

A small loop antenna is a very inefficient radiator. Its radiation resistance is very small in comparison with its loss resistance, and both are dominated by its self inductance. Thus, the loop electrically appears to be an inductor. The value of this inductance depends on its size and shape. A circular loop in air with radius much larger than wire diameter follows the simple law[8]:

$$L = \mu_o a[\ln \frac{16a}{d} - 2] \tag{2}$$

where L is the loop inductance, μ_o is the premeability of free space, a is the loop radius, and d is the wire diameter. The current flowing in this loop is related to the voltage applied by the obvious relation:

$$I = \frac{V}{\omega L} \tag{3}$$

From these considerations, it will be seen that the voltage necessary to produce a given field depends strongly on the size of the loop. The dipole strength, IA, varies directly with loop area,

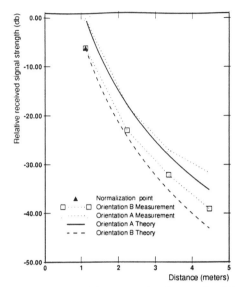

Figure 6: Comparison of the theoretical free-space signal strength with the measured values for two loop antennas at a height of 0.35m above a soggy field. Orientations A and B of Figure 5 are shown.

while the inductance varies only as the square root of area times the logarithm of the area. The dipole moment then is an increasing function of loop area, when the loop is driven from a constant voltage source. That is, a given voltage will result in a greater field as the area of the loop is increased.

The loop also acts as a receiving antenna. Similar to a transformer secondary, the induced voltage is given by:

$$V_{rec} = j\omega\mu_o H_n A \tag{4}$$

where V_{rec} is the induced voltage, and H_n is the component of magnetic field strength normal to the loop. The loop inductance, L, and loss resistance then appear in series with this induced voltage. In general, the receiver antenna design is not very critical in the HF band, since the noise floor is primarily determined by EMI[9].

The desirable isolation effects of near-field coupling can be further enhanced through the use of a modulation technique which exhibits capture effect. Frequency modulation schemes [9][10] such as FSK are simple to implement, and exhibit a high degree of capture for situations in which signal-to-interference ratios are greater than 10db. Thus, cell isolation can be further improved by the use of FSK.

3   Hardware Implementation

The Nano-Cellular LAN consists of fixed stations (base stations) and portable stations. Figure 3 schematically depicts the system in which fixed stations define the cells, and portable stations are scattered among the cells. Fixed stations consist of a 68302-based processor connected to a

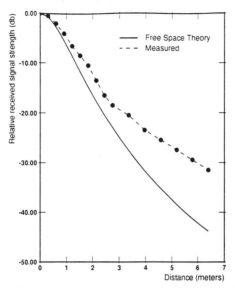

Figure 7: Comparison of the theoretical free-space signal strength with the measured value for a loop receive antenna receiver 1.2 m above the floor and a quadrature transmitter on the ceiling in an open area inside the building. The measurement corresponds to orientation D of Figure 5 and Geometry A of Figure 8.

conventional LAN (10Mbit Ethernet) and equipped with a near-field transceiver. Portable stations consist of MPADs[11] equipped with near-field transceivers.

The most important design features of the near-field transceivers are associated with the loop antenna. From the considerations of the previous section it can be determined that for a transmitter with 5 volt power supply, a loop of area .0244 m^2 (assuming a wire diameter of 3mm) will meet the FCC requirement. Loop antennas commonly used for UHF television reception have an area of approximately .025 m^2 and an inductance of 550nH. The portable station antennas have been implemented with these loops. Base station antennas are similar, but have one notable distinction. A single loop antenna will exhibit a strong orientation dependence. In order to make the system somewhat orientation independent, base station antennas each consist of two loops perpendicular to each other, and driven in phase quadrature. This results in a circularly polarized field with much better angular uniformity than that obtained from a single loop. Reception from this structure is also performed in phase quadrature, so that the reception pattern is similarly affected. Were it not for the larger volume of this structure, the portable stations could be similarly equipped.

Block diagrams for the two types of transceivers are shown in Figure 4. Aside from the quadrature circuitry, the designs are identical. These transceivers operate in a half-duplex mode. The circuit consists of a transmitter, a receiver, and a transmit/receive (T/R) switch.

The transmitter circuitry is extremely simple. It is based on a Colpitts oscillator with two possible operating frequencies. This frequency agility is accomplished through switching circuitry which allows the resonator capacitance to change between two values. The oscillator operates at either 20.4Mhz or 22.0Mhz. A digital divider then reduces this frequency by a factor of four to give the operating carrier of 5.3Mhz with a deviation of ± 200khz. The divider also generates the quadrature signals necessary for base station use. The oscillator and divider are implemented

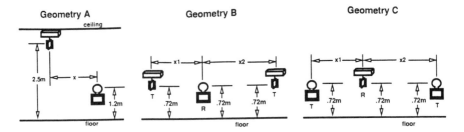

Figure 8: Geometry A: A portable station communicating with a ceiling mounted base station. Geometry B: Two base stations transmitting to the same portable receiver, all at a height of 0.72 meters above the floor. Geometry C: Two portable devices transmitting to the same base station, all at a height of 0.72 meters above the floor.

with 74HC series components. Divider output then drives a class-C power output stage based on a 2N3906 PNP transistor which directly drives the loop antenna. The loop antenna is provided with appropriate parallel capacitance (2500-3000pf) for resonance at the center of the 5.3Mhz channel.

The receiver circuitry is based on an FM IF/quadrature detector integrated circuit (Signetics NE614). The carrier frequency of 5.3Mhz is well within the range over which this device can provide adequate gain and detection. Thus, the device is used as a tuned radio frequency (TRF) receiver with no frequency conversion. A simple LC bandpass filter is provided for selectivity. All receiver circuitry has a moderate Q, of approximately 10, so that no crystal control is necessary. The quadrature detector operates well at this frequency, and can deliver demodulated data at a rate of 250kbps. A simple analog comparator follows the quadrature detector to convert the demodulator output to digital levels and to add a small amount of hysteresis.

The T/R switch consists of a CD4007 MOS transistor array configured to isolate the receiver from the loop during times when the transmitter is operating. Without this feature, the receiver will experience a shift in its bias points and, hence, a reduction in gain, in the presence of the transmit signal. The time required to recover from this effect can seriously affect system performance. The T/R switch serves to greatly improve the receiver's recovery time. It has been found that only a few bit times are required for recovery.

Data recovery is performed by a state machine implemented in a Xilinx part. This state machine performs clock recovery by sensing transitions of the received data signal, and samples the data at the midpoint between transitions. Data is then processed by the HDLC controller of the 68302.

4 Hardware Performance

In order to make a comparison with theory, an initial set of measurements were made in an open field, using loop antennas in the two orientations shown in Figure 5. Figure 6 shows a comparison between the measured field strength and the theoretical free-space value for two orientations of receive and transmit antennas. Only the relative amplitude of the measured points was determined, so the comparison with theory is normalized to make the indicated point agree. In these measurements, signal strengths were always at least 10db above the noise floor.

The measured and theoretical curves both exhibit a 6dB difference between the two orientations, as is expected from the field equations[12]. At large distances between transmitter and receiver, the measured and theoretical values diverge significantly. This discrepancy is most likely due to the

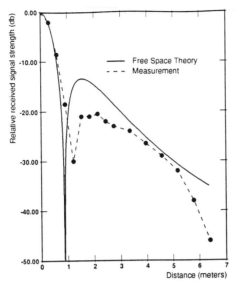

Figure 9: Comparison between theoretical and measured signal strength for a a ceiling mounted transmitter and a loop antenna receiver as a function of horizontal separation. The measurement corresponds to orientation C of Figure 5 and Geometry A of Figure 8.

fact that the earth acts as a weak conductor, and excludes magnetic flux lines. At 5 Mhz, the skin depth for moist soil is on the order of 3 m [13][14]. With the antennas situated only 0.35 m above the soggy ground, one would expect to see deviations from free-space behavior when the distance between antennas is on the order of one skin depth. As can be seen in the Figure 6, deviations from free-space behavior do in fact become prominent at a spacing of about 3 m.

In an indoor environment, the 5 MHz signals are significantly distorted and attenuated by the metal in furniture and walls. Figure 7 shows the measured signal strength in an open area inside Xerox PARC. The measurement corresponds to orientation D of Figure 5 and geometry A of Figure 8, with a ceiling mounted quadrature antenna acting as transmitter and a loop receive antenna located 1.2 m above the floor. The deviation from free-space theory is much more pronounced than for the outdoor result.

For Orientation C and Geometry A, there is a fairly non-intuitive null in the received signal strength which occurs some horizontal distance from the transmitter and for a narrow range of receiver antenna orientations. This null is due to the use of a single loop antenna in the receiving apparatus, and can be predicted from the field equations. The null region actually has a conical shape, such that its horizontal position depends upon the vertical position. The effect is shown in Figure 9, where the measured result exhibits a strong null at a horizontal distance of 1.2 m. The slight difference between the measured and theoretical position of this null is due to distortions of the field caused by the metal floor and wall framing.

Figure 10 shows the signal strength for both antenna orientations as a function of position in the presence of an intervening wall. The signal is sharply reduced as the receiver traverses the wall, an effect due to shielding that is provided by the metal studs used to frame the walls.

The metal structures in the building act to change the magnitude and direction of the signals. This can be seen in Figure 11, which shows the dependence of received signal strength on antenna

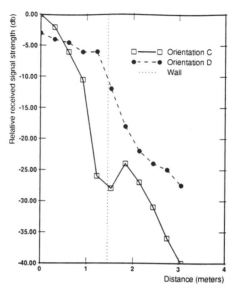

Figure 10: Received signal strength for orientations C and D of Figure 5 as the receiver traverses a wall between offices. The wall is indicated by the vertical dotted line. The plot shows results for Geometry A of Figure 8.

orientation for various transmitter-receiver spacings. Figure 11(a) shows the theoretical free-space result, while (b) shows the measured result in the middle of an open area in the building. As can be seen in (b), the direction of maximum field strength has been strongly altered by the building. The angular extent of the Orientation C null can be seen in the theoretical curve for a separation of 0.91 m.

From an installation standpoint, it would be preferable to mount base stations on the ceiling. However, the presence of a strong null in received signal strength at a horizontal distance of approximately one meter makes this geometry highly problematic. In addition, we have discovered that electronic ballasts used for fluorescent lighting can interfere with base station reception, and reduce the range to less than 1 m. These problems are not present when the base stations and portable devices are all at the same height above the floor. The remainder of our measurments were therefore made with base stations and portables at a height of 0.72 m above the floor, as shown by Geometries B and C of Figure 8.

Radio reception is governed by the signal-to-noise ratio at the receiving antenna. In order to characterize the noise, a number of measurements were made. The measurement apparatus consisted of a simple loop antenna, tuned for resonance at 5.3Mhz with a Q of 10, a linear low-noise amplifier, and a spectrum analyzer. The spectrum analyzer was set to have an IF bandwidth approximately equal to that of the actual receivers. The loop probe was then used to observe the time-dependent noise within a region of the PARC building.

The noise in this frequency band tends to be dominated by man-made interference. This type of EMI is particularly prevalent in an indoor environment. Several characteristics of the noise are apparent. The highest amplitude contribution to the noise consists of relatively infrequent (intervals of tens of msec) impulses. The lower amplitude contributions also consist of impulses, but at a higher repetition rate. At certain orientations, the noise consists of regular bursts of pulses at a very

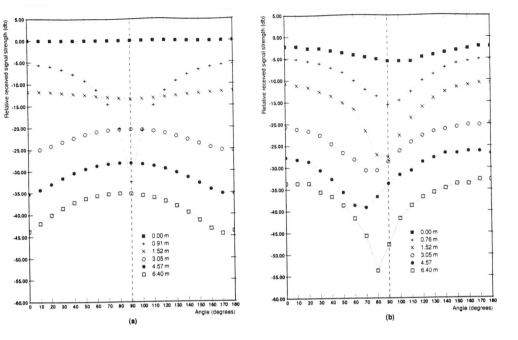

Figure 11: (a) Theoretical dependence of received signal strength on receiver antenna angle for various distances from the transmitter. The midpoint of the ordinate corresponds to orientation C of Figure 5. (b) The measured values in an open area of the building. Both plots corresponds to Geometry A of Figure 8.

low frequency of 30 Hz (not synchronous with the 60 Hz line frequency.) Attempts to correlate these impulses to specific pieces of equipment (motors, etc) in the building were inconclusive. Additional measurements revealed that the noise amplitude can change by large amounts (10db) over distances on the order of one meter.

The reception of data packets depends on the signal-to-noise ratio at a given location. It is generally accepted that for FSK systems, noise which is more than 10 dB below the signal strength does not interfere with data reception. Figure 12 shows the fraction of received packets for a loop transmitter and quadrature receiver at a height of 0.72 m above the floor, and for orientations C and D of Figure 5. The range for reliable reception of more than 90% of the packets is approximately 2.5 m.

The optimal positioning of base stations depends upon multiple factors, including the instrinsic range of the channel in a given noise environment, and on the performance of the system when multiple mobile devices are operating in close proximity to one another. Ideally, the physical positioning of base stations and the communication protocol will be tuned to minimize the detrimental impact of physical regions in which such interference occurs.

One situation in which interference can occur corresponds to geometry B of Figure 8 and orientation C of Figure 5, where two fixed base stations are transmitting at the same time. Shown in the Figure 13 is the measured fraction of received packets as the receiver is moved between the transmitters. There is a finite range over which the two transmitters interfere with reception, leading to a complete loss in packet reception. On either side of this region, the receiver accurately

152

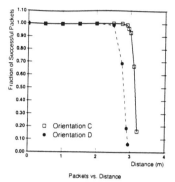

Packets vs. Distance

Figure 12: Fraction of successful packets as a function of distance for a loop transmitter and quadrature receiver, both 0.72 m above the floor. The results correspond to orientations C and D of Figure 5.

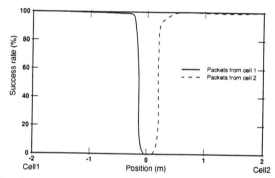

Figure 13: Fraction of successful packets as the receiver is moved between two transmitters which are at a fixed separation of 4 m. The result corresponds to geometry B of Figure 8 and orientation C of Figure 5.

senses packets from the nearest transmitter.

Figure 14 shows a more general result, corresponding to geometry C of Figure 8. In this case, two portables, at distances x1 and x2 from a base station, are transmitting packets at the same time. Plotted in the Figure 14 are lines of constant percent received packets. The contour lines diverge at larger values of x, implying that the range over which interference can occur becomes larger as the separation between transmitters increases.

For typical applications in office buildings, it will be reasonable to distribute base stations so that there is one cell per office. The near-field system is particularly well suited to this application, since the instrinsic cell radius is approximately 2.5 m, and since the signals are significantly attenuated by walls. For applications in open areas of a building, it will be desirable to distribute base stations in such a way that the boundaries between cells are located at physical boundaries (partitions, bookshelves, etc.) In general, the optimum spacing between base stations will be on the order of $2 * 2.5 = 5.0$m.

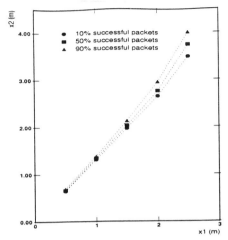

Figure 14: Contours of constant percentage packet reception as the locations of two mobile transmitters are varied independently. The result corresponds to geometry C of Figure 8 and orientation C of Figure 5.

5   Media Access Protocol

In a wireless LAN, carrier and collision are not global properties as they are in a traditional broadcast LAN, such as Ethernet. The stations work around the absence of carrier sense and collision detection by exchanging a pair of short control packets (RTS and CTS) before sending a data packet. After a successful exchange of control packets, with high probability all relevant stations know that a data packet is about to be sent and can avoid collisions. To our knowledge this technique was first suggested for wireless media access control by Karn [15], where it is called MACA for "Media Access with Collision Avoidance."

We were initially concerned that our radios would have a high error rate. Although this has not turned out to be as serious as was feared, it is true that the most likely place for a packet to be dropped is the wireless link. Moreover, the protection afforded to a DATA packet by the RTS-CTS exchange is considerably reduced if there are multiple, overlapping cells. For these reasons we took the obvious step of adding acknowledgements to MACA. Now data is sent between stations A and B by a 4-way exchange:

1. Station A sends RTS to station B, indicating its desire to send data.

2. Station B replies CTS, indicating its willingness to accept the data.

3. Station A sends data.

4. Station B sends ACK.

Stations not directly participating in the exchange (called "bystanders") must refrain from sending in order not to disrupt the exchange. They do so according to the following rules:

- Rule D1: A bystander hearing an RTS packet must defer for the time required to send the expected CTS packet.

- Rule D2: A bystander hearing a CTS packet must defer for the time required to send the expected DATA packet.

• Rule D3: A bystander hearing a DATA packet must defer for the expected ACK packet.

Intuitively, these rules are justified as follows. For each pair of stations (A, B), we make the "reciprocity assumption" that A can hear transmissions from B iff B can hear transmissions from A; i.e., there is a symmetric notion of A being in range of B. It follows that packets sent by A can collide with packets being received by B iff A can hear transmissions from B. Thus, if A hears an RTS sent by B, A should defer to avoid colliding with the expected CTS at B (rule D1). Similarly, if A hears a CTS sent by B, A should defer to avoid colliding with the expected DATA packet at B (rule D2). However, if A hears an RTS sent from B to C but does not hear an answering CTS, then with high probability C either has failed to respond or is out of range of A. In either case A may legitimately send packets without disrupting communication between B and C. Thus, there is no requirement that a bystander hearing an RTS packet defer for the expected DATA packet.

The last rule (D3) does not protect ACK packets to the extent that rules D1 and D2 protect CTS and DATA packets. To see this, consider three stations A, B and C, where A is within range of B and B is within range of C, but A and C are out of range. Suppose B sends an RTS to C (which is also heard by A) and B receives an answering CTS (not heard by A). While B sends its DATA, A may legitimately engage in communication with some other station. Thus, a packet sent by A can collide with the ACK eventually sent by C to B. Fortunately, such a collision is not especially costly. After a link-layer timeout (on the order of a single contention slot), B resends the RTS. Since the data has already been transferred successfully, C responds with an ACK, which (like a CTS) is protected by rule D1. Thus, while some additional contention is introduced, the data is not re-sent, and the extra overhead is independent of the size of the data packet.

Stations contend for the medium by sending RTS packets until one of them successfully elicits a CTS response. Given the above defer rules, is easy to see that nearly all collisions occur between RTS packets or between an RTS and a CTS packet. A successful CTS indicates the medium has been acquired. We use a fairly standard exponential backoff strategy[11].

In practice we cannot assume every mobile station will hear every successful exchange to adjust its backoff parameter. Thus, the stations' backoff parameters will not track one another precisely. This can lead to asymmetric situations in which a successful sender may reduce its backoff and contend more aggressively than other stations, allowing it to dominate the channel for long periods. Although this behavior keeps channel utilization high, it is particularly undesirable for our system, in which most communication involves interactive use of hand-held devices.

Our solution to this problem is to reserve a byte in the header of each MACA packet for the current backoff value. Whenever a packet is received, the receiving station adopts the backoff value from that packet. Since every station in a cell is within range of the cell base station, and all communication involves the cell base station, it follows that the most recently successful backoff value is used with high probability by every station in the cell, avoiding the unfair behavior described above. This has been verified by simulation (up to 100 stations per cell) and experiment (5 stations per cell).

The technique of copying backoff values achieves fairness only in the sense that every station wishing to acquire the channel has an equal chance of doing so. This definition of fairness is clearly inadequate for most cellular networks, in which about half the traffic in each cell originates at a single site (the cell's base station). In our current architecture, virtually all traffic comprises a single bidirectional TCP stream per portable machine, connecting the halves of the X server. To deal with this problem, our MACA code maintains a separate output queue for each destination, and behaves as if it were running a separate MACA output process for each destination. The behavior of the real station is governed by the most aggressive of the simulates output processes.

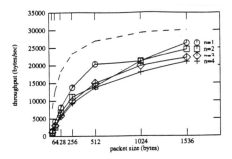 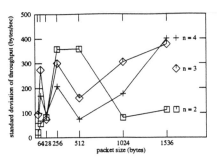

Figure 15: MACA throughput performance with fixed length packets

6 Measured System Performance

To assess the system performance of a multi station Nano-Cellular network, we performed a series of measurements using MACA to provide reliability and fair access to the medium. The measured configuration was a single cell containing a base station and from 1 to 4 portable stations.

Each portable was configured to act as an infinite-capacity source of data packets directed at the base station. Software on the base station logged each packet exchange, sending packet trace data across the Ethernet to a file server for analysis off-line. The volume of trace data was only a few percent of the base station's Ethernet capacity. We varied the number of portables and the size of the data packets. All data packets in an individual test were of uniform size. Each test was several minutes long.

The left hand graph in Figure 15 plots total throughput as a function of packet size and number of portables. The dotted line represents a theoretical maximum data capacity, obtained by computing the total size of MACA packet headers, HDLC framing and CRC overhead for a four-way MACA exchange as described in the previous section, assuming no packet turnaround time.

Notice that the data rate increases with increasing packet size. This is because for larger packets there are fewer packets per second, and so less time is lost to collisions. Notice also that, for a fixed packet size, the data rate decreases with increasing number of stations, since there are more collisions per packet. The difference between the measured throughput and the theoretical maximum is due to collisions and to the nonzero packet turnaround time, which dominate for small packet sizes.

The right hand graph in Figure 15 plots the standard deviation of the throughput for the various tests. In all cases the value is a small fraction of the measured throughput. We do not yet understand the marked dip for 128-byte packets.

7 Summary

The unique features of near-field coupling have been exploited to build a nanocellular wireless network. The benefits of this technique include cell sizes appropriate to individual offices, very simple and low cost hardware, and sufficient bandwidth for many applications. A well-known media access protocol (MACA) was modified to address the unique constraints of a limited range, common channel wireless medium. The custom protocol has very low overhead, and the net system bandwidth is sufficient to implement a custom X server across the radio link.

References

[1] Terri Watson and Brian Bershad. Local area mobile computing on stock hardware and mostly stock software. In *Proceedings of the USENIX Mobile & Location-Independent Computing Symposium*, pages 109–116. USENIX Association, 1993.

[2] N. Adams, R. Gold, B. Schilit, M. Tso, and R. Want. An infrared network for mobile computers. In *Proceedings of the USENIX Mobile & Location-Independent Computing Symposium*, pages 41–51. USENIX Association, 1993.

[3] R. Want and A. Hopper. Active badges and personal interactive computing objects. *IEEE Transactions on Consumer Electronics*, 38(1):10–20, 1992.

[4] K. Pahlavan. Wireless intraoffice networks. *ACM Transactions on Office Information Systems*, 6(3):277–302, July 1988.

[5] Theodore S. Rappaport. Indoor radio communications for factories of the future. *IEEE Communications Magazine*, pages 15–23, May 1989.

[6] Hiroshi Shigeno, Kaname Arai, Teruo Yokoyama, and Yutaka Matsushita. A hybrid indoor data network with radio and wire - radio propagation measurements and performance evaluation in a rayleigh channel. In *Proceedings of the 16th Conference on Local Computer Networks*, pages 276–281. IEEE Comput. Soc. Press, 1991.

[7] Federal Communications Commission. Title 47 - telecommunication. In *Code of Federal Regulations*. U.S. Government Printing Office, 1989.

[8] S. Ramo and J. R. Whinnery. *Fields and Waves in Modern Radio*. Wiley, 1944.

[9] U. L. Rohde and T. T. N. Bucher. *Communications Receivers*. McGraw-Hill, 1988.

[10] Henry Stark and Franz B. Tuteur. *Modern Electrical Communications*, chapter 7, pages 347–348. Prentice-Hall, 1979.

[11] Christopher A. Kantarjiev, Alan Demers, Ron Frederick, and Robert T. Krivacic. Experiences with x in a wireless environment. In *Proceedings of the USENIX Mobile & Location-Independent Computing Symposium*, pages 117–128. USENIX Association, 1993. also avaiable as Xerox PARC Technical Report CSL-93-9.

[12] Roger F. Harrington. *Time Harmonic Electromagnetic Fields*. McGraw-Hill, 1961.

[13] P. Lorrain and D. Corson. *Electromagnetic Fields and Waves*. W.H. Freeman and Company, 1970.

[14] *Reference Data for Radio Engineers*. Howard W. Sams, 1975.

[15] Phil Karn. Maca - a new channel access method for packet radio. In *Proceedings of the ARRL 9th Computer Networking Conference, London Ontario, Canada*. ARRL, September 1990.

11

Radiowave Propagation Measurements for Sharing Spectrum Between Point-to-Point Microwave Radios and Personal Communications Systems

S.Y. Seidel, D.M.J. Devasirvatham, R.R. Murray, H.W. Arnold, and L.G. Sutliff

Bell Communications Research
331 Newman Springs Road
Red Bank, NJ 07701

Abstract

A CW radiowave propagation experiment was performed in the New York City region to study spectrum sharing possibilities between point-to-point microwave systems and personal communications systems (PCS). The results showed that the isolation measured between the microwave and a PCS system on the street was more dependent upon building shadowing and street orientation within the city than on the distance between the two systems for distances less than 20 km. At greater distances, the separation between transmitter and receiver causes a significant increase in isolation due to increased terrain variations and earth curvature. The system dependent isolation required for sharing the spectrum with a few PCS radiators could be found even in the conventional main beam direction in regions shadowed by surrounding objects. Conversely, strong signals were often found in directions well away from the main beam. Thus, there was measurable degradation of the microwave antenna discrimination relative to its "free-space" directivity. At all distances, the isolation varied widely, with noticably strong signals popping up in larger shadowed areas. Low power systems with low antenna heights and small cell sizes, on the order of a city block or less, would have a better probability of finding suitably isolated regions to operate in. Large cell, high-powered systems with high antennas may need to move well beyond the horizon to share spectrum.

1. Introduction

Frequencies in the 1.9 GHz band have recently been allocated for emerging Portable Communications Systems around the world. In the United States and the United Kingdom, these frequencies have been granted under the proviso that the spectrum should be shared between the PCS services and incumbent point-to-point microwave communications systems. Permissible levels of interference into point-to-point microwave systems are governed by industry practices. In the US, these are detailed in Bulletin 10, published by the Telecommunications Industry Association (TIA), which, in its latest revision, Bulletin 10-F has attempted to formulate uniform practices for both the Common Carrier and the Private Microwave sectors of the industry. This paper details a radiowave propagation experiment which was performed to provide information on spectrum sharing possibilities for Bulletin 10-F. It studies the reception on the ground of signals from a test point-to-point microwave antenna. The results are presented in a manner meaningful to both the point-to-point microwave and PCS provider community.

2. The Experiment

A four-foot (1.2 m) parabolic grid antenna was used to represent one end of a point-to-point microwave link, operating in the 1850-1990 MHz band. The transmitter power was +34 dBm. The nominal antenna gain was 26 dBi, giving an equivalent isotropic radiated power (EIRP) of +60 dBm. The receiver system was in a pickup truck. The vertically polarized receiver antenna was mounted at a height of 5.8 ft. (1.8 m). It had dipole-like pattern with a nominal gain of 0 dBi. The receiver bandwidth was only 1 kHz. The detected signal was continuously sampled at intervals of about 1.8 inches (4.6 cm) of travel. Thus there were about three samples per wavelength of the signals. The vehicle was also equipped with a non-satellite based, map matching, geolocation system which enabled the vehicles position to be known at all times.

The transmitter was placed, successively, at two sites in Manhattan, New York City. The first site was the Empire State Building, at 34th Street and 5th Avenue. The NYNEX radio room on the 87th floor was used.

This site was 1076 ft. (328 m) above ground level. The antenna room was on the eastern face of the building, with the radome-like wall extending partially to the northern and southern faces. One end of several operational microwave links were also located in this room. The antenna was pointed "north" along 5th Avenue, at an azimuth of 29 degrees for two sets of data. This is also the approximate orientation of Manhattan island. The main beam extends into upstate New York. The transmitting antenna was vertically polarized for the first set and horizontally polarized for the second. For the third set of data, the antenna was pointed "south" at Wall Street, at an azimuth of 203 degrees. The main beam of this vertically polarized antenna crossed the bay into New Jersey. The fourth data set was obtained with the microwave antenna placed outside on a parapet, atop the NYNEX building at 210 West 18th Street. The antenna was pointed "south" at Wall Street. Any radiation from the back of the antenna was blocked by structures of the building. The azimuth was 194 degrees and the antenna was vertically polarized.

The geolocation was obtained from the measurement equipment along with the data consisting, usually, of 1000 signal samples (155 ft., 48 m in total distance), and stored on disk. The vehicle was driven on the avenues and streets of the city and on the highways outside, traveling within a radius of about 120 km. All the runs made with each pointing of the microwave antenna made up a set that consisted of several hundreds of miles (km) of data. A complete calibration of the measurement system was performed before undertaking the measurement campaign.

Since the transmitter radiated a CW signal, the received data were subject to multipath fading fluctuations. Hence the median isolation of each complete 1000 sample record was extracted, and the geolocation, which was originally recorded at the end of each record, was recalculated to the center of the record. This spatial processing over about 48 m was considered to characterize the signal well over the given distance.

3. Isolation

3.1 Path Loss and Isolation

The generalized Friis' free space transmission formula between two antennas may be written, in logarithmic form, as

$$P_r = P_t + U_t(\phi) + U_r(\varphi) - L_p \text{ [dB]} \tag{3.1-1}$$

where,

P_t = Transmitter Power in dBm
P_r = Received Power in dBm
$U_t(\phi)$ = Gain Function of transmitter antenna in the direction ϕ of the receiver in dBi
$U_r(\varphi)$ = Gain Function of receiver antenna in the direction φ of the transmitter in dBi
L_p = Path Loss between Isotropic antennas in dB

Path loss will *always* be defined between isotropic antennas and antenna gains referred to the isotrope as well, whether the qualifier is used or not. We may further write the Gain Function of an antenna as

$$U(\phi) = G - D(\phi) \text{ [dB]} \tag{3.1-2}$$

where

G = Gain of the antenna = $U(\phi)_{max}$ is the maximum value of the Gain Function.
$D(\phi)$ = Discrimination function of the antenna, in the direction ϕ.

The value of the discrimination function in the direction of maximum gain is 0 dB. Hence, any antenna may be thought of as in isotropic radiator with gain, G, which is enclosed in a shell with attenuation in different directions described by the discrimination function, D(ϕ).

The generalized transmission formula (3.1-1) can be rewritten in logarithmic form as,

$$P_r = P_t + G_t + G_r - (L_p + D_t(\phi) + D_r(\phi)) \text{ [dBm]} \qquad (3.1\text{-}3)$$

We define the Isolation, L_I in dB as

$$L_I = L_p + D_t(\phi) + D_r(\phi) \text{ [dB]} \qquad (3.1\text{-}4)$$

so that

$$P_r = P_t + G_t + G_r - L_I \text{ [dBm]} \qquad (3.1\text{-}5)$$

and

$$P_r = \text{EIRP} + G_r - L_I \text{ [dBm]} \qquad (3.1\text{-}5a)$$

The Isolation includes the path loss and the discriminations of the two antennas in a link. The Isolation equals the path loss only when isotropic antennas are used, or when antennas are aligned along their directions of maximum gain (usually along the boresight in the case of a parabolic antenna). During measurements, it allows the antenna gain, or alternatively, the Equivalent Isotropic Radiated Power (EIRP) to be used in the transmission formula. Since the three-dimensional angular discrimination function is usually not known, its effect is lumped into the Isolation term. We see, therefore, that during signal level measurements using antennas with gain, when the mobile receiver is at some arbitrary position relative to the microwave antenna, the only propagation loss quantity which can be measured directly is the Isolation.

3.2   Required Isolation for Spectrum Sharing

TIA Bulletin 10 requires that when a PCS link is installed at a frequency used by point-to-point microwave links, the interference generated by the PCS system should not raise the thermal noise floor of an existing microwave link by more than 1 dB. Therefore, since the desired and interfering powers are uncorrelated, the power received from the PCS system, if it is noise-like, should be at least 6 dB below the microwave radio's noise floor [1]. Table 3.5-1 shows the interference budget for a typical microwave receiver.

TABLE 3.2-1: Co-channel interference budget for a microwave receiver.

Parameter	Value	Noise Contribution
Thermal noise / Hz		-174 dBm
Effective noise bandwidth	10 MHz	+70 dB
Noise Figure	4 dB	+4 dB
Total Thermal Noise		-100 dBm
Maximum permissible interference + thermal noise	1 dB degradation	-99 dBm
Hence, maximum permissible interference power P_int_max (after power addition)		-106 dBm

Thus, the total interference permitted from a coordinating PCS system is only -106 dBm. The isolation required between a PCS system and the microwave radio to achieve this may be calculated according to the following two examples. Consider two types of portable sets, operating in time division multiplexed systems with 8 bursts per TDMA frame. The burst power when on is assumed to be 1 watt for one set which is vehicle based and 100 mw for the other, which is a hand portable. The portable antenna gain, is assumed to be 3 dB for the vehicular unit and 0 dB for the hand portable. Their base stations, in each case, are assumed to transmit continuously at 1 watt, using omni-azimuthal base antennas with gains of 10 dB each. Now consider a microwave radio at that frequency which has an antenna gain of 30 dB.

TABLE 3.2-2: Required co-channel isolation between a PCS system and a microwave receiver.

Parameter	Base Station	Vehicular System	Hand Portables
PCS transmitter burst power when on (P$_t$)	30 dBm	30 dBm	20 dBm
Duty factor with 8 time slots per frame	0 dB	-9 dB	-9 dB
PCS antenna gain (G$_t$)	10 dB	3 dB	0 dB
Microwave receiver antenna gain (G$_{MW}$)	30 dB	30 dB	30 dB
Maximum permissible interference power, P_int_max, from Table 3.5-1	-106 dBm	-106 dBm	-106 dBm
Hence, minimum required isolation for a single unit to share spectrum (A)	176 dB	160 dB	147 dB
Minimum required isolation for each of 100 units to share spectrum (B)	196 dB	180 dB	167 dB
Minimum required isolation for each of 1000 units to share spectrum (C)	206 dB	190 dB	177 dB

Note that this table does not imply that all three classes of emitter are co-existing with the same microwave radio simultaneously. Each class is being evaluated by itself. It does show that higher power PCS systems require more isolation to avoid violating the TIA sharing criterion. Also, more users could be supported if some of them have greater isolation than the minimum numbers given above. Conversely, if a single unit were to move to a position with less isolation than the numbers given above in rows B or C, a large number of users would have to increase their isolation significantly to maintain the same total number of users. Even this can only continue as long as the user with the lowest isolation does not reach the single user value, (A), given in Table 3.5-2. It should also be noted that the TDMA duty factor correction is strictly applicable only for PCS interferers spread uniformly over all time slots.

4. Measured Data

In this section, we show examples of measured data. To provide perspective to the data, we relate the measured isolation to the system examples given in Section 3.2. The isolation levels given in the examples were selected to fall within the range of measured data. They are for illustrative purposes only and do not imply that PCS systems can be economically deployed for the corresponding numbers of users given in Table 3.2-2. The reader must repeat the calculation of isolation levels that correspond to the appropriate number of users for actual PCS systems.

4.1 Measured Isolation

Figure 1 shows a scatter plot of Isolation (dB) vs. T-R Separation (km) for co-polarized measurement locations with the transmitter on the Empire State Building (328 m high) with an Azimuth of 29 degrees. Each data point represents the median value over a distance of about 155 feet. Notice the large variability of isolation for a fixed T-R separation. The figure indicates several locations at distances less than 20 km where there is sufficient isolation for a few PCS radiators to share the spectrum. However, in order to guarantee that there are no locations within a given distance that do not have sufficient isolation, a PCS radiator must be located over the horizon (>80 km) from a sensitive microwave receiver. Figure 2 shows the cumulative distribution of isolation for all locations less than 20 km from the transmitter. The four curves in Figure 2 represent measured data from different transmitter locations (physical separation, azimuth, or polarization).

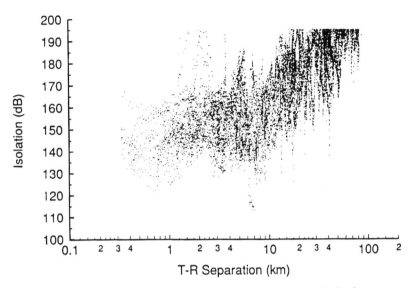

Empire State Building: Azimuth=29 degrees
Polarization/Height: Tx-Vertical/328 m, Rx-Vertical/1.7 m

Figure 1 Scatter plot of Isolation (dB) vs. T-R Separation (km) for co-polarized measurement locations with the transmitter on the Empire State Building (328 m high) with an Azimuth of 29 degrees.

Consider an isolation of 180 dB for a high power vehicular PCS system in the example given in Table 3.2-2, column 2. If each user is at 180 dB isolation, at most 100 users in the whole area can share the spectrum. Figure 2 shows that users located less than 20 km from the microwave receiver have 180 dB or greater isolation at 10-25 percent of the locations.

Next, as an example, consider the low power system described in the 3rd column of Table 3.2-2. Figure 3 shows a plot of all measurement locations for the Empire State Building transmitter location with an azimuth of 29 degrees and vertical transmitter polarization. Figure 3a shows the locations where the isolation is greater than 180 dB, and Figure 3b shows the locations where isolation is less than 180 dB. The transmitter is located at (0,0), and the solid lines represent a five degree beam along the antenna boresight. Each "dot" on the graph is the median isolation computed over 155 feet (1000 points).

The region shown includes parts of southern New York State, Westchester, and Connecticut to the north, the I-95 highway along the coast to the northeast, Long Island, to the east, and parts of New Jersey to the west and south, with Sandy Hook bay in between. The beam passes over Westchester, which has moderately hilly terrain, covered with vegetation. It is immediately seen from Figure 3a that isolation at the 180 dB level exists even on the highway very close the beam for this reason. However, there are a significant number of off-boresight locations, even over 30 km away, which do not have this isolation. These are found interspersed among the locations which exceed this isolation. Conversely, close in at Long Island, there are regions of high isolation in the midst of areas of high signal strength.

Figure 4 shows regions above and below the 180 dB isolation level in Manhattan Island when the transmitter was at West 18th street. Pockets of isolation, not very large in extent, with at least 180 dB isolation in the

162

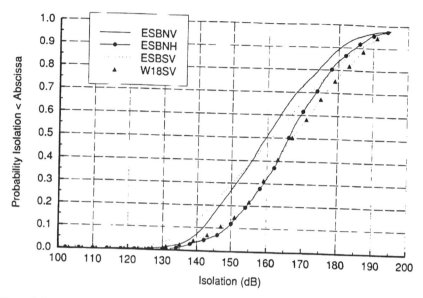

Figure 2 Cumulative distribution of isolation for all measured locations less than 20 km from the transmitter. Each curve represents a different set of measurement data.

rear of the antenna can still be seen. However, the occasional sections of relatively low isolation in these regions would need to be excluded to ensure that the total interference does not exceed the allowable level (see Table 3.2-1).

These data show the difficulties of simply using a geographic exclusion zone method for spectrum sharing. Usable regions in Westchester would have been discarded under this scheme. On the other hand, even if more of these usable zones could be identified with some other method, there is always the danger of a user wandering into an "isolation hole." This would produce excessive interference into a microwave system unless some form of monitoring is done [2]. The data also shows that small-cell systems would have a better opportunity to co-exist in this environment.

4.2 Antenna Front-to-Back Ratio

At the Empire State Building, measurements were recorded at two transmitter antenna azimuths, 29 and 203 degrees. Although not quite the full 180 degree front-to-back separation, measurements from this antenna combination can be used to estimate the realizable front-to-back ratio of the antenna. Figure 6 shows measured isolation for receiver locations between 24 and 34 degrees azimuth. This figure depicts measurements that are in the "main beam" of the 29 degree transmitter azimuth and measurements in the "back lobe" of the 203 degree transmitter azimuth. Regression lines are shown for the two sets of isolation. The distance dependence slopes are nearly identical with about a 20 dB offset between the two lines. This gives us a preliminary estimate of 20 dB for the realizable front-to-back ratio of the antenna measured at ground level. The manufacturer lists the front-to-back ratio of the antenna as 30 dB. The 10 dB degradation in front-to-back ratio is caused by multipath scattering that has "filled-in" the area behind the transmitter and perhaps also by the receiver not being on the boresight plane. Hence, published microwave antenna discrimination

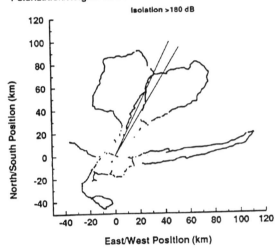

Empire State Building: Azimuth=29 degrees

Polarization/Height: Tx-Vertical/328 m, Rx-Vertical/1.7 m

Isolation >180 dB

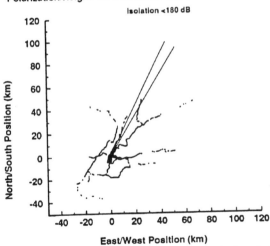

Empire State Building: Azimuth=29 degrees

Polarization/Height: Tx-Vertical/328 m, Rx-Vertical/1.7 m

Isolation <180 dB

Figure 3 Drive-around plot of measurement locations with the vertically polarized transmitter on the Empire State Building with an Azimuth of 29 degrees. The top plot (a.) shows the measurement locations where measured isolation was greater than 180 dB, and the bottom plot (b.) shows locations where the isolation was less than 180 dB.

164

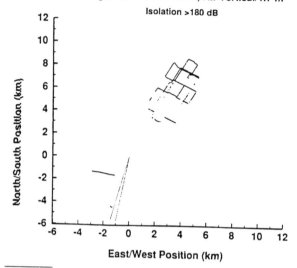

West 18th Street: Azimuth=194 degrees
Polarization/Height: Tx-Vertical/93 m, Rx-Vertical/1.7 m
Isolation >180 dB

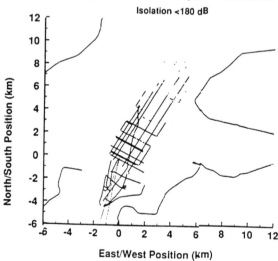

West 18th Street: Azimuth=194 degrees
Polarization/Height: Tx-Vertical/93 m, Rx-Vertical/1.7 m
Isolation <180 dB

Figure 4 Drive-around plot of measurement locations with the vertically polarized transmitter on the NYNEX Telephone Building. The top plot (a.) shows the measurement locations where measured isolation was greater than 180 dB, and the bottom plot (b.) shows locations where the isolation was less than 180 dB.

factors must be used with caution in computing the obtainable isolation to a PCS system, even from a very high location.

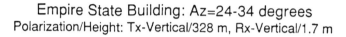

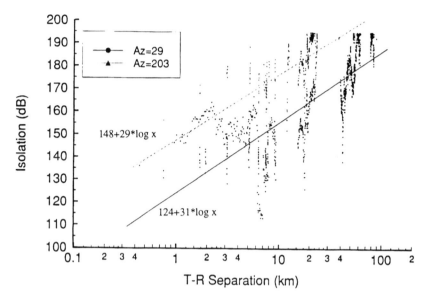

Figure 5 Scatter plot of measured isolation (dB) for receiver locations between 24 and 34 degrees Azimuth with the transmitter on the Empire State building. The different symbols indicate whether the transmitter boresight was at 29 or 203 degrees Azimuth. Also shown are the best-fit regression lines for each set of measurement data.

4.3 Discussion

The results underline the differences between the comparatively well behaved microwave radio terrain propagation environment and the rapid short-scale variation of the PCS world. At short distances, he terrain and the clutter in which the PCS system is embedded become very strong factors and dominate the isolation. The terrain and clutter also contribute to the large variability in isolation at a given distance. The separation between the transmitter and receiver only introduces significant isolation at very long distances when the earth bulge begins to intrude. A complex area such as New York City has some regularity in its plan layout, even though it may be very irregular in the third dimension. Thus, close in, it may be amenable to modeling if building height data were available. Further away, the terrain along the line-of-sight and the surface cover information may give good results, with the city appearing as a clutter factor, causing additional attenuation.

5. Summary

A radiowave propagation experiment was performed in the New York City region to study spectrum sharing possibilities. A CW transmitter radiated from a parabolic antenna that was placed successively in two buildings where a point-to-point microwave link could be used. The received signal was measured from a vehicle which traveled in New York, New Jersey, and Connecticut. The results showed that building shadowing and street orientation within the city had a stronger impact than the distance between the two systems for distances less than 20 km on the measured isolation between the microwave and a PCS system on the street. At greater distances, the separation between transmitter and receiver causes a significant increase in isolation due to increased terrain variations and earth curvature. The results showed that for short distances, the isolation measured between the microwave system and a personal communications system on the street depended only weakly on the distance. It depended strongly on the building shadowing and street orientation within the city. At greater distances, the topography, including intervening hills and the curvature of the earth which could shield the systems, played a strong role in providing isolation. The system dependent isolation required for sharing the spectrum with a few PCS emitters, which significantly exceed the free space value, could be found even in the conventional main beam direction in locations shadowed by the surrounding objects. Conversely, strong signals were often found well away from the main beam. Thus, there was measurable degradation of the microwave antenna discrimination. Closer in, the isolation varied widely, with regions of strong signals popping up in larger shadowed areas. Hence, low power systems with low antenna heights and small cell sizes, on the order of a city block or less, would have a better probability of finding suitably isolated regions to operate in. Large cell, high-powered systems with high antennas may need to move well beyond the horizon to share spectrum. It should be emphasized, however, that the minimum required isolation for sharing is system and capacity dependent and not related to distance in any way. The measurement data also show that antenna direction discrimination cannot be depended on to increase the isolation between point-to-point microwave and personal communications systems.

6. References

[1]   V.K. Varma, et. al., "Interference, Sensitivity, and Capacity Calculations for Measurement-based Wireless Access Spectrum Sharing," *43rd IEEE Vehicular Technology Conference*, Secaucus, NJ, May 1993, pp. 550-554.

[2]   V.K. Varma, et. al., "A Beacon Detection Method for Sharing Spectrum Between Wireless Access Systems and Fixed Microwave Systems," *43rd IEEE Vehicular Technology Conference*, Secaucus, NJ, May 1993, pp. 555-559.

12

Estimating Regions of Service in Wireless Indoor Communication Systems

Paul M. Cartwright and Kevin W. Sowerby

Department of Electrical and Electronic Engineering, University of Auckland

Auckland, New Zealand.

email: pm.cartwright@auckland.ac.nz, kw.sowerby@auckland.ac.nz

ABSTRACT

Future indoor wireless communications systems are likely to be cellular in structure with very small cells, employing frequency reuse both horizontally and vertically. This paper investigates the effects that signal variability, mean power and obstacles in the indoor environment may have on system performance. Results show that if the desired signal suffers Rician fading, the spectral efficiency is higher than that for Rayleigh fading. Walls in buildings increase the path loss between cochannel cells allowing a reduction in the reuse distance and a concomitant increase in the spectral efficiency. Noise emanating from office equipment is likely to influence system performance in indoor environments and a technique for treating noise as stochastic interference is presented.

1. INTRODUCTION

It is envisaged that wireless technology will soon permit fast, efficient and seamless communication between users virtually irrespective of their locations. This paper discusses the possible performance of wireless communication systems specifically designed to operate within buildings. In this paper, system performance is described in terms of the reliability of reception, or *outage probability*.

Since only a small band of spectrum is likely to be allocated for a particular wireless indoor service, extensive frequency reuse within buildings will probably be required to support large numbers of users. Indeed, in an indoor environment, frequencies may have to be reused both horizontally and vertically (as shown in Fig. 1).

Cochannel interference (CCI) is an inherent and undesirable property of any frequency reuse system. Within a building CCI could potentially be a more serious problem than in two-dimensional outdoor systems because cochannel base stations may be distributed in three dimensions. Because CCI is likely to limit the capacity of indoor systems, understanding and estimating its effects in indoor environments is essential if system designs are to be effective and efficient. There are two general classes of CCI, namely *intra-building* and *inter-building* interference. Inter-building CCI emanates from nearby buildings with similar wireless services, while intra-building CCI is that which emanates from the building in which the receiver is located. The primary focus of this paper is the effects of intra-building CCI on system performance.

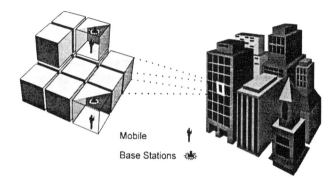

Fig. 1. A built-up area with base stations located in three dimensions providing a ubiquitous communications service.

Propagation models for fading and mean power in indoor channels are described in Section 2. Section 3 presents performance measures, namely spectral efficiency and outage probability which are used to illustrate the effects of intra-building CCI. In indoor environments, walls and the strength of the line-of-sight signal affect system performance. These effects are illustrated in the results presented in Section 4.

Two techniques for treating noise as either cochannel interference or as a minimum signal requirement are presented in Section 3. Section 4 presents results of how noise can degrade reception quality in a small-cell indoor environment

2. PROPAGATION MODELS

2.1. FADING STATISTICS

There are various probability density functions (pdfs) that describe the variability of the received signal strength in indoor environments. In this paper two pdfs are used, namely the Rayleigh and Rician distributions. The Rayleigh distribution (or Rayleigh fading) is a useful model where no dominant multipath component exists [1]. In the indoor systems modelled in this paper, cochannel cells are assumed to be spaced sufficiently far apart for their signal amplitudes in a nearby cochannel cell to be Rayleigh distributed. Accordingly their powers are distributed exponentially and the pdf of the ith cochannel interferer, $f(y_i)$, is given by

$$f(y_i) = \frac{1}{Y_i}\exp\left(-\frac{y_i}{Y_i}\right), \qquad \qquad ...(1)$$

where Y_i is the mean power of the ith interferer. In this paper, the letter y denotes an interfering signal and x denotes the desired signal.

The Rician distribution is an appropriate model of signal variability where a dominant (or line-of-sight, LOS) component and a number of weaker multipath components are present [1]. The Rician distribution is used in this paper to model the variability of the desired signal at the point of reception. The pdf of the desired signal power conditional on its mean power, $f(x|X)$, is given by

$$f(x|X) = \frac{1}{X'} \exp\left(-\frac{2x+s^2}{2X'}\right) I_0\left(\frac{\sqrt{2xs}}{X'}\right),$$
...(2)

where X' is the average scattered power, $s^2/2$ is contributed by the LOS component and the total mean power is, $X = X' + s^2/2$. The Rician factor, $a = s^2/2X'$, is the ratio of the LOS power to the scattered power. When there is no LOS component (i.e. $s = 0$ and $a = 0$), Eq. 2 reduces to the exponential distribution in Eq. 1.

The relationship between the scattered power (X'), mean power (X) and Rician factor (a) is straightforward but justifies explanation as various authors define these variables differently. It can be shown that if the scattered power is related to the total mean received power by a scaling factor, namely $X' = X/(1+a)$ then $X' + s^2/2$ will always equal the total mean power. If this scaling is performed in an analysis, then increasing the Rician factor increases the relative intensity of the LOS component while maintaining the total mean power constant [2]. If this scaling is not performed then the overall mean power will increase with increasing Rician factor [3]. This latter result can be optimistic since for a particular Rician factor, the mean power at a location might end up higher than that predicted by a mean power propagation model.

2.2. MEAN POWER PREDICTION

Reception reliability is influenced by the variabilities of the received signals and their mean powers. The estimation of the mean power of a signal in indoor channels can be performed using a number of models. The simplest model is that commonly used for large cell outdoor environments, where the mean power, $\overline{P}_r$, received at a mobile is a function of the transmitter-receiver (Tx-Rx) separation, d, namely [4]

$$\overline{P}_r = \frac{k}{d^\gamma},$$
...(3)

where γ is known as the propagation exponent and k incorporates the effects of quantities that do not depend on the propagation path, such as the antenna gains and transmitter power. In indoor environments, attenuation for short Tx-Rx separation (d) is often close to free space but as d increases, the rate of attenuation also increases as the signal propagates through a greater number of walls. Eq. 3 implies a constant rate of attenuation and consequently it does not describe mean received power accurately over a large range of separations. Despite this, Eq. 3 represents an attractive model for mean power prediction due to its computational simplicity. Reported values for γ in indoor environments vary considerably - ranging from 1 for LOS paths up to 6.5 for highly obstructed paths [1]. It is interesting to note that when γ is less than 2, the mean power is higher than that predicted by the free space model. La Fortune et al report that this primarily occurs in corridors where a wave canalisation effect is observed [5].

A more complex model for predicting mean power incorporates a free space path loss for d and includes an additional dB/(unit distance) attenuation, ρ, namely [6]

$$\overline{P}_r = k \frac{10^{-(\rho/10)d}}{d^2}. \qquad \qquad ...(4)$$

Typical values of ρ are between 0.3dB/m and 1.2dB/m [1].

A more location specific model for mean power prediction incorporates a free space path loss for d and additional discrete attenuations for each wall and floor that obstruct the direct propagation path. This model can be expressed in logarithmic form as [7],

$$\overline{P}_r(dB) = 10\log(k) - 20\log(d) - FAF \cdot m - WAF \cdot n, \qquad ...(5)$$

where WAF and FAF are the wall and floor attenuation factors respectively and m and n are the number of floors and walls in the propagation path. Although Eq. 5 has a relatively simple form it can be difficult to apply to situations of practical interest. This is because estimation of $\overline{P}_r$ at a location in a building requires the application of a geometric test to determine how many walls and floors are present in the propagation path.

A comparison of the mean power models in Eqs. 3, 4 and 5 is presented in Fig. 2. It can be seen that all three models give similar mean powers for short Tx-Rx separations.

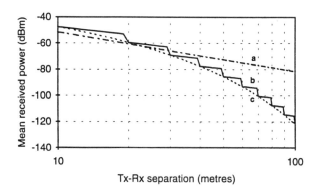

Fig. 2 Mean received power as a function of Tx-Rx separation for three propagation models. (a) Eq. 3 with $\gamma = 3$, (b) Eq. 5 with $WAF = 6$dB/wall and walls spaced at 10m intervals. (c) Eq. 4 with $\rho = 0.6$dB/m.

However as d increases, curve "a" in Fig. 2 (which is calculated from Eq. 3) does not show an increased rate of attenuation thus giving optimistic values for $\overline{P}_r$. Curve "c" (which is calculated from Eq. 5) shows discrete changes in the mean power after each wall in the propagation path. In this example, the walls are assumed to be spaced evenly at 10m intervals with each wall providing 6dB of attenuation. Curve "*b*" (which is calculated from Eq. 4) follows the average slope of curve "*c*". This would not occur if the walls were separated by random distances.

The accuracy of these three models in indoor environments is uncertain. Even Eq. 5 does not model the effect of walls accurately as it is reported in [5] that the attenuation associated with a wall (i.e. the *WAF*) is not constant but depends on how far the receiver is located from the wall. These models for mean power do not take into account electromagnetic phenomena such as reflection and diffraction. However Eqs. 4 and 5 do allow for an increased rate of attenuation as a function of Tx-Rx separation.

3. MEASURES OF SYSTEM PERFORMANCE

To compare the effects that parameters of indoor environments have on system performance, two measures are defined, namely *reception quality* and *spectral efficiency*. It is important to realise that spectral efficiency must be considered in association with the quality of reception because the simultaneous maximisation of quality and efficiency is not usually possible.

3.1.   SPECTRAL EFFICIENCY

Spectral efficiency is a measure of how effectively a system uses its allocated frequency resources to provide a specific service. A simple but useful measure of spectral efficiency is the number of channels provided per cell. (This particular measure is also called *radio capacity* [8]). The measure *channels per cell* is determined by dividing the total number of unique channels available to the system by the required cluster size. The advantages of frequency reuse in a cellular design are reflected by this simple measure. Its particular value will depend on the frequency reuse distance, the tolerable level of CCI and the total number of channels available in the allocated spectrum.

3.2.   OUTAGE PROBABILITY

Outage probability can be defined as: "the percentage of time that the quality of reception is unacceptable due to cochannel interference and noise". Generally minimum acceptable carrier-to-noise (CNR) and carrier-to-interference (CIR) ratios are specified for a particular modulation scheme.

3.3.   OUTAGE PROBABILITY IN THE PRESENCE OF COCHANNEL INTERFERENCE

For situations where the desired and interfering signals suffer Rayleigh fading and the "interference only" case is considered, the outage probability, P_{out}, is given by [9],

$$P_{out} = 1 - \prod_{i=1}^{I} \left(\frac{\Lambda_i}{\alpha + \Lambda_i} \right),\qquad \dots(6)$$

where I is the number of active cochannel interferers and $\Lambda_i = X / Y_i$ is the CIR for the ith cochannel interferer. The protection ratio, α, represents the minimum CIR for acceptable reception quality and P_{out} is the percentage of time that the actual CIR is less than α.

For situations where the statistics of the cochannel signals are described by Rayleigh fading and the desired signal is described by Rician statistics, the probability of outage, P_{out}, for the "interference only" case is given by [3],

$$P_{out} = 1 - \sum_{i=1}^{I} \left[1 - \frac{\alpha}{\alpha + \Lambda_i'} \exp\left(-\frac{a\Lambda_i'}{\alpha + \Lambda_i'} \right) \right] \times \prod_{j=1, j \neq i}^{I} \frac{\Lambda_j'}{\Lambda_j' - \Lambda_i'} , \qquad \ldots(7)$$

where $a = s^2 / 2X'$ is commonly called the Rician factor. The scattered power to interference ratio for the kth interferer is given by $\Lambda_k' = X'/Y_k$. As explained in Section 2.1, to maintain the mean power constant while adjusting the Rician factor it is required that $\Lambda_k' = \Lambda_k /(1+a)$. As expected, if $a = 0$, (i.e. there is no LOS component) then Eq. 7 reduces to Eq. 6.

Implementation of Eq. 7 is complicated by the fact that it suffers from numerical instability if any of the carrier to interference ratios (Λ_k) are identical. Eq. 7 has been derived specifically assuming that all of the CIRs are different. However by slightly perturbing CIR values which are equal, Eq. 7 can still be applied without introducing any significant error.

3.4. OUTAGE PROBABILITY IN THE PRESENCE OF NOISE AND INTERFERENCE

Section 3.3 presented expressions for outage probability in the presence of cochannel interference only. This section presents outage probability expressions for environments in which both noise and cochannel interference are present.

Thermal noise in receivers is always present, however outages due to this kind of noise can usually be minimised by increasing transmitter powers (or by implementing some kind of diversity reception). Indoor wireless systems will be deployed in environments laden with electromagnetic devices in close proximity to both base stations and mobile units. Equipment such as computers, CRTs, elevators, TVs, microwave ovens, etc, are often found in office buildings and may cause interference to a communications link [10]. Noise (mainly thermal) has been treated in previous outage probability expressions but usually the noise power is assumed to be constant and is not treated as cochannel interference. Under these conditions, P_{out} for Rayleigh distributed desired and interfering signals can be written as [9]

$$P_{out} = 1 - \exp\left(\frac{-r_n}{\Gamma_N} \right) + \sum_{i=1}^{I} \frac{\alpha}{\alpha + \Lambda_i} \exp\left\{ \frac{-r_n}{\Gamma_N} \left(\frac{\alpha + \Lambda_i}{\alpha} \right) \right\} \prod_{j=1, j \neq i}^{I} \frac{\Lambda_j}{\Lambda_j - \Lambda_i}, \qquad \ldots(8)$$

where Γ_N is the carrier to noise ratio (CNR) and r_n is the noise protection ratio. All other symbols are as defined previously. If the noise power is treated like CCI and is also assumed to vary exponentially then P_{out} is given by [11],

$$P_{out} = 1 - \left(\frac{\Gamma_N}{\Gamma_N + r_n} \right) \prod_{i=1}^{I} \left(\frac{\Lambda_i}{\alpha + \Lambda_i} \right). \qquad \ldots(9)$$

Eq. 9 can be written in its elegant form only because the noise has been assumed to have the same form of pdf as the cochannel interferers. Allowing variability in the noise power and treating it as a form of cochannel interference allows its effects to be estimated more precisely (see [11] for more details). For the analyses involving noise in this paper, the CNR $\left(\Gamma_N\right)$ is defined as $\Gamma_N = \beta\varepsilon / d^2$, where ε incorporates the effects of the antennas and the free space constants, d is the Tx-Rx separation and β is the ratio between the transmitter power and the mean noise power.

4. SYSTEM EVALUATION

4.1.   SYSTEM DESCRIPTION AND BUILDING LAYOUT

In order to investigate the effects that features of the indoor environment have on system performance, a mathematical model of a building is required. The building model adopted in this paper consists of a stack of identical floors. The plan for each floor is shown in Fig. 3. Each room is 10 metres square, i.e. an area of $100\,\mathrm{m}^2$. Floors are assumed to be spaced 5m apart and are stacked directly on top of each other. The overall dimensions of the building are not specified as they do not affect the results in this analysis. Each 10x10m room is assumed to be a cell with its own base station positioned centrally on the ceiling. Base station antenna patterns are assumed to be omni-directional in the horizontal plane but are directed from downwards from the ceiling to the floor. Cochannel cells are separated by the reuse distance D.

If a fixed channel assignment strategy is employed then, with reference to the square cell layout in Fig. 3, each cell could be surrounded by up to eight closest cochannel cells on the same floor. Only these eight dominant interferers are included in the estimation of outage probability.

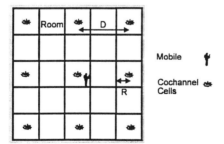

Fig. 3   An arbitrary arrangement of cochannel cells on a floor in a building. Each room is a cell with radius $R = 5$m. The reuse distance, D, in this example is 20m.

The indoor system modelled in this paper is assumed to employ TDMA modulation with 8 channels multiplexed into a 200kHz channel. If 2MHz of spectrum space is allocated to the entire indoor service, then 40 unique duplex channels would be available to allocate throughout the building. To estimate the spectral efficiency, the required cluster size is determined and the forty channels are then shared evenly amongst the cells in the cluster. The interference protection ratio, α, is maintained at 8dB, which is an appropriate value for second generation digital systems [12].

4.2. EFFECT OF SIGNAL VARIABILITY AND WALLS

The following analysis investigates how the LOS component of the desired signal affects system performance. Fig. 4 shows the effect of varying the Rician factor (a) in a free space environment. The results presented are for the situation where eight cochannel interferers are located around a given base station and are separated by the reuse distance D. Mean signal powers can be predicted using Eq. 3 with $\gamma = 2$. For each Rician factor the minimum reuse distance has been determined such that the outage probabilities in all locations of a 10x10m room are less than 5%. It can be seen that for Rayleigh fading ($a = 0$ in Fig. 4), the required reuse distance, D, is close to 180m. As the Rician factor increases to 10 the required reuse distance drops to only 70m. As a comparison, the effect of a solitary cochannel interferer located at distance D is also shown in Fig. 4.

This analysis is particularly appropriate to buildings which have long corridors. In these environments, propagation may be close to that of free space. In other building environments, the free space model may be somewhat unrealistic, however the results from this analysis show a marked reduction in the minimum required reuse distance as the Rician factor is increased.

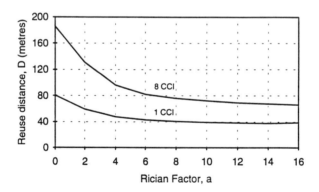

Fig. 4 Required reuse distance versus Rician factor in a free space environment. The two curves are for 8 cochannel interferers and one cochannel interferer.

Fig. 5 plots the required wall attenuation factor (*WAF*) as a function of Rician factor (a). In this case walls are included in the propagation environment and spaced 10m apart forming rooms with dimensions of 10x10m. Accordingly, Eq. 5 is used to estimate mean powers. For a given spectral efficiency (channels per cell) and Rician factor, the minimum *WAF* that yields outage probabilities less than 5% in all locations of a room has been determined.

For example, a cluster size of 4 (10 channels/cell) requires 9dB/wall under Rayleigh fading to meet the 5% reception quality criterion but only 5dB/wall if the desired signal has a Rician factor of 10. Typical values for *WAF* are around 6dB [1] and typical Rician factors in indoor environments are also around 6 [13]. Thus with reference to Fig. 5, practical systems might be designed with approximately 4 cells per cluster.

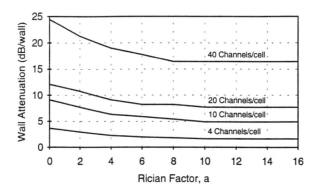

Fig. 5   Required wall attenuation factor versus Rician factor for a range of spectral efficiencies. The *WAF* is chosen so that all locations in a 10x10m room have outage probabilities less than 5%.

The interference resulting from cochannel cells on different floors of a building and the effects of floor attenuation factors are discussed in [14].

4.3.   EFFECT OF NOISE

Fig. 6 shows the effect of outage probability versus Tx-Rx separation. In this analysis, eight cochannel interferers, a cluster size of 4 and wall attenuation of 9dB/wall have been assumed. Curve "*a*" in Fig. 6 is calculated from Eq. 9 and is for exponentially distributed signal powers, namely the desired signal, the interfering signals and the noise. It has been assumed that $r_n = 12\text{dB}$ and $\beta = 70\text{dB}$. Curve "*b*" in Fig. 6 is calculated from Eq. 8 and is for exponentially distributed desired and interfering signals and constant noise power that is not added to the interference ($r_n = 12\text{dB}$ and $\beta = 70\text{dB}$). Curve "*c*" in Fig. 6 is calculated from Eq. 6 and is the interference only case i.e. $r_n = 0$. Note the sharp transition in outage probability at $d = 5\text{m}$. This corresponds to the wall at the edge of the room. At this point, the mean power of the desired signal drops by the *WAF* and the mean power of some of the interferering signals may increase by the *WAF*. As expected, including a variable noise power in the total interference (curve "*a*") yields a higher outage probability than that which results when the effects of noise are included only as a minimum signal requirement (curve "*b*").

If $r_n = 12\text{dB}$, then a margin, β of 90dB, between the transmitter power and the noise floor corresponds to the CCI interference limited situation. For this case the outage probability as a function of distance is shown as curve "*c*" of Fig. 6. If for example, the thermal noise floor is -110dBm then a transmitter power of only -20dBm would correspond to this 90dB margin. However if noise other than thermal noise is considered then its mean power may be considerably higher than -110dBm and its effects may need to be included in the estimation of system performance.

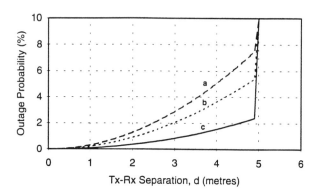

Fig. 6 Outage probability versus Tx-Rx separation for various noise and interference situations. (a) CCI with exponentially varying noise, Eq. 9. (b) CCI + effects of noise included by using a minimum signal threshold, Eq. 8. (c) CCI only, Eq. 6.

5. CONCLUSIONS

The capacity of future indoor wireless communications systems is likely to be limited by cochannel interference since frequency reuse may be employed both horizontally and vertically. Consequently, the way in which propagation in the indoor channel influences system performance needs to be understood.

Propagation in indoor channels is primarily due to the multi-path effect, however signal variability may be less severe than Rayleigh fading since line-of-sight paths are likely in small cell environments. Under line-of-sight conditions, the desired signal may be assumed to have Rician statistics. It is shown in this paper that in small-cell indoor systems, Rician fading yields an improvement in spectral efficiency compared to Rayleigh fading.

Walls and floors within buildings increase the path loss between cochannel cells thus allowing smaller reuse distances. Buildings with wall attenuation factors close to 5dB/wall and values for the Rician factors for the desired signal of around 6 should be able to support cluster sizes between 5 and 8.

Typical transmitter powers in small-cell systems may be large enough to overcome the effects of thermal noise. However other forms of "noise like" interference found within office buildings may have mean powers significantly higher than thermal noise. It is shown that treating noise like cochannel interference results in a lower estimate of the reception quality compared to that obtained when noise is treated as a minimum signal requirement.

6. ACKNOWLEDGMENTS

The authors gratefully acknowledge the support of *Telecom Corporation of New Zealand Ltd*.

REFERENCES

[1]   HASHEMI, H., 'The indoor radio propagation channel', *Proc. IEEE*, Vol. 81, No. 7, Jul. 1993, pp. 943-968.

[2]   PRASAD, R. and KEGEL, A., 'Performance of microcellular mobile radio in a cochannel interference, natural and man-made noise environment', *IEEE Trans. Veh. Tech.*, Vol. 42, No. 1, Feb. 1993, pp. 33-40.

[3]   YAO, Y., and SHEIKH, A., 'Outage probability analysis for microcell mobile radio systems with co-channel interferers in rician/rayleigh fading environment', *Electron. Lett.*, Vol. 26, 1990, pp. 864-866.

[4]   LEE, W. C. Y., 'Elements of cellular mobile radio systems', *IEEE Trans. Veh. Tech.*, Vol. VT-35, No. 2, May 1986, pp. 48-56.

[5]   LAFORTUNE, J., and LECOURS, M., 'Measurement and modeling of propagation losses in a building at 900MHz', *IEEE Trans. Veh. Tech.*, Vol. 39, No. 2, May 1990, pp 101-108.

[6]   PORTER, P. T., 'Relationships for three-dimensional modelling of co-channel reuse', *IEEE Trans. Veh. Tech.*, Vol. VT-34, No. 2, May 1985, pp. 63-68.

[7]   SEIDEL S., et al., The impact of surrounding buildings on propagation for wireless in-building personal communications system design', *IEEE 42nd Vehic. Tech. Conf.*, Denver, 1992, pp. 814-818.

[8]   LEE, W. C. Y., 'Spectrum efficiency in cellular', *IEEE Trans. Veh. Tech.*, Vol. 38, No. 2, May 1989, pp. 69-75.

[9]   SOWERBY, K. W., WILLIAMSON, A. G., 'Outage probability calculations for multiple cochannel interferers in cellular mobile radio systems, *IEE Proc.*, Vol. 135, Pt. F, No. 3, June 1988, pp. 208-215.

[10]   BLACKARD, K. et al, 'Measurements and models of radio frquency impulsive noise for indoor wireless communications', *IEEE J Selected Areas in Commun.*, Vol. 11, No. 7, Sep. 1993, pp 991-1001.

[11]   GURR, H., SOWERBY, K. and WILLIAMSON, A., 'Generalised outage probability and BER estimation using dual protection margins', to appear in *IEEE 44th Vehic. Tech. Conf.*, Stockholm, 1994.

[12]   BEZLER, M. *et al*, 'Comparison of spectrum efficiency in digital cellular systems: GSM and AMPS-D', *IEEE 42nd Vehic. Tech. Conf.*, Denver, 1992, pp. 1008-1011.

[13]   BULTITUDE, R., 'Measurement, characterisation and modelling of indoor 800/900 MHz radio channels for digital communications', *IEEE Comms Mag.*, Vol. 25, No. 6, Jun. 1987, pp 5-12.

[14]   CARTWRIGHT, P. and SOWERBY, K., 'Reception Reliability in Three Dimensional Personal Communications Systems', to be published in *Proc. IEEE Vehicular Technology Conferernce* 1994, Stockholm, Sweden.

13

Adaptive Propagation Prediction using Lee's Model in a Non-Homogeneous Environment

Robert Lopez
e-mail: rob@ram.net

Konstantinos Vlahodimitropulos
e-mail: vlahodim@rcc.com

RAM Communications Consultants, Inc.
10 Woodbridge Center Dr.,
Woodbridge, NJ 07095

Abstract

The need to improve the accuracy of signal propagation prediction in an environment of variable building/clutter density drove the development of a tool that would facilitate the compilation and analysis of field collected signal strength measurements to improve results of future radio coverage analyses. The output of the analysis produces a set of signal attenuation parameters that are representative of the environment tested. The Lee Area-to-Area Path Loss Prediction model was chosen as the methodology with which the measured field parameters were applied. These derived parameters were the 1-mile intercept point and path-loss slope particular to Lee's model. The approach for deriving these parameters is based on measuring an unmodulated carrier and correlating the measurement sample with a GPS derived location. The collection of measurements was grouped by the environment classification, for example, dense urban, urban, suburban, etc. Therefore, each of these environments are represented by a new set of Lee parameters. The adaptive or automatic assignment of these parameters is facilitated with the use of a spatial land-use database which contains information about the environment classification areas. This database enables the propagation characteristics of the environment to be automatically changed as a radio unit traverses from area to area. In the case where the propagated signal crosses environment boundaries, special care is taken to adjust the rate of signal attenuation using Lee's paradigm.

Introduction

Conventional methods of performing propagation predictions involve the use of a general purpose electromagnetic wave expression as a function of distance, frequency, antenna gains, antenna heights, etc. where received signal levels are further adjusted by losses due to terrain variations. Changes in environment surrounding a base station are usually accounted by adding a net loss value derived from experience and/or published studies for a specific set of conditions. Research results ([1],[2]) have shown that these adjustments cannot be done by simply applying a correction value in dBto the results of the propagation prediction model. A more sophisticated method is proposed by Lee [2], which suggests changing dynamically the loss equation throughout the area in question. Applying this propagation model can be done through the use of a Geographic Information System (GIS) which includes spatial databases, such as terrain elevations as well as land-use classifications, i.e., forestation, hydrography and building density. The terrain elevation database is used mostly for modeling terrain shadows and diffraction. The land-use database is used to adjust the propagation loss parameters as they are represented in the propagation prediction model.

In this paper the methodology proposed by W.C. Lee for the treatment of non-homogeneous environments and the way it is applied with a GIS are described. An overview of the method for collecting and processing signal strength measurements is given, as well as the way these measurements are fed back to the propagation prediction system.

Lee Model Application

The Lee model [4] includes a path loss equation together with knife edge diffraction. For the knife edge diffraction computation, a terrain elevation database is used to generate the terrain profile of the propagation path, and the diffraction equation is applied to the profile parameters.

This work is primarily focused on the path loss components which can be uniquely sensitive to the propagation environment.The path loss equation of the Lee model has the form:

$$P_r = P_{r0} - \gamma \log(r) - n \log(f) + \alpha \qquad \text{(EQ 1)}$$

where P_r is the received power, P_{r0} is the **1 mile intercept**, γ is the **path loss slope**, r is the distance from the site, f is the frequency and α is a factor of antenna heights, transmitted power, etc.

The two highlighted parameters, intercept and slope, are derived from field strength measurements. Thus, this model gives the flexibility to adapt to heterogeneous environments by using a library of known (intercept, slope) pairs for various environment types.

EQ 1 has the property that it can allow the system to use more than one value of γ, when the path connecting the base station site and the mobile spans over different environment types. Special care has also been taken to calculate the 1 mile intercept in the case when, within the 1 mile radius from the site, the environment changes.

The Terrain Land-Use Database

This database contains the characteristics of the environment, such as hydrology and building density. This information is used in models that can adapt to changing environment even if the change occurs within a single propagation path. This information is usually derived from satellite imagery. Better than 25 meter precision data is available.

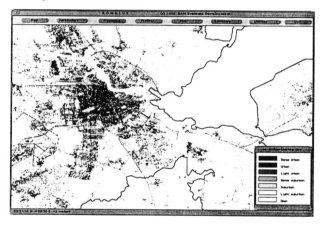

FIGURE 1. A view of the land-use data for the area of Amsterdam.

The application of the Lee model utilizes these data as described in the previous section. Specifically, the slope γ is adjusted every time the propagation path crosses a boundary between two types of environment. Moreover, the 1 mile intercept is calculated as the average intercept attribute of the environment types within the first mile from the site. Thus, the second term at the right hand side of EQ 1, is broken into a sequence of terms, each one corresponding to each environment type occurring between the path connecting the site and the mobile, as shown in the formula:

$$P_r = \overline{P}_{r0} - \gamma_1 \log(r_1) - \gamma_2 \log(r_2/r_1) - \gamma_3 \log(r_3/r_2) \ldots - n \log(f) + \alpha \qquad \textbf{(EQ 2)}$$

where $\overline{P}_{r0}$ is the averaged intercept, γ_n and r_n are the slope attribute of the nth type of environment that was met along the propagation path and r_n is the distance from the site at which this environment type is met.

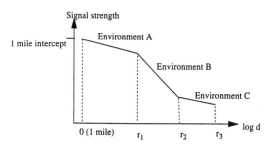

FIGURE 2. Received signal strength due to path loss vs. log(distance).

For efficiency reasons, this information is organized in grid cells, that can be accessed individually or as group, when the application queries the environment types along a propagation path or within 1 mile.

Measurement Analysis and Comparison

Signal strength measurements are an essential part of Radio Planning, because they give the ability to the user to verify the predictions made by the application of a Propagation Prediction model. They also help the engineer adjust the environmental losses of a propagation model to accurately predict coverage in the widely diverse areas of the world..

Obtaining Local Average Power

The measurements are done using a spectrum analyzer and are georeferenced using a GPS receiver. Special care has been taken to comply with Lee's specifications [2]. The sampling frequency was about 10 measurements per second.

The measurement sequences were split into sweeps of 1 second length. From each sweep the median is extracted. Assuming that, within 1 second, the variation due to the lognormal fading is negligible, these median values have lognormal distribution only. The effects of the fast fading (Rayleigh) component are minimized by taking these medians. The medians are then arranged spatially, in a grid containing 500m by 500m cells. The values at each grid cell are averaged in order to obtain values for the received power as a function of distance from the site.

Studies by RAM in the Den Haag area in Holland have shown that the RMS difference between the mean and median of 422 sweeps containing a total of 216,064 samples is less than 1.3 dB. It is reasonable to assume that the local median as measured approximates the local mean [2].

FIGURE 3. Measurements vs. Predictions for Den Haag, Netherlands.

Also, by fitting these medians to a log-linear curve, the 1 Km (or 1 mile) intercept and the attenuation slope are obtained. If the measurements are taken in a homogeneous area (with respect to land-use), these values can be used as descriptive parameters of the environment type (EQ 2).

Comparison with Predictions

Using the grid structure mentioned in the previous section, a propagation model is used to estimate the received power at each grid cell. Then, each predicted value is compared with the average measured value of each cell. The difference between the predicted value and the averaged measured value is averaged to obtain the level of optimism (or pessimism) of the predictions with respect to the measurements. The standard deviation of the difference can be used in the calculation of the prediction reliability.

Conclusions

This paper has presented the use of Lee's model to automatically adapt the path loss slope and 1 mile intercept pairs to heterogeneous environment typically surrounding a base station site. Using field measurement data to derive new Lee parameters for the area under test and a high resolution land-use database, the bridge between actual and expected results may be narrowed by implementing a systematic and efficient comparative process.

184

More work has to be done in the correlation of environment characteristics with signal attenuation in order to achieve more efficient modelling of the radio environment using high precision land-use databases.

Acknowledgments

The authors thank Mr. Richard Sadler of RAM Communications Consultants Ltd. for performing the signal strength field measurements in The Netherlands, Mr. Michael Hunter of RAM Communications Consultants, Inc. for his suggestions and Mr. Sadhan Mandal for his support.

References

[1] W.C.Y. Lee, Mobile Communications Design Fundamentals., Howard W. Sams & Co., 1986.

[2] W.C.Y. Lee, Estimate of Local Average Power of a Mobile Radio Signal, IEEE Trans. on Vehicular Technology, Vol. VT-34, No. 1, Feb. 1985.

[3] Y. Okumura, et. al. Field strength and Its Variability in VHF and UHF Land-Mobile Radio Service, Review of the Electrical Communication Laboratory, Vol. 16, 9-10, Sept.-Oct. 1968.

[4] M. Hata, Empirical Formula for Propagation Loss in Land Mobile Radio Services., IEEE Trans. on Vehicular Technology, Vol. VT-29, No. 3, Aug. 1980.

[5] K. Bullington, Radio Propagation for Vehicular Communications. IEEE Trans. on Vehicular Technology, Vol. VT-26, No. 4, Nov. 1977.

[6] R.B. Carey, Technical Factors Affecting the Assignment of Facilities in the Domestic Public Land Mobile Radio Service, FCC Research Division, Report No. R-6406.

[7] H.P. Stern, S. Darbha, An Adaptive Propagation Program, University of Alabama, Technical Report presented to RAM Broadcasting Corp., Sept. 1992.

[8] K. Vlahodimitropulos, A. Metaxas, ROMULUS™ User's Manual, RAM/BSE Communications, L.P., 1993.

14

An Interactive System for Visualizing
Wireless Communication Coverage within Buildings

Manish A. Panjwani, A. Lynn Abbott, Theodore S. Rappaport,
Virginia Polytechnic Institute and State University
The Bradley Department of Electrical Engineering
Blacksburg, Virginia, USA

and

Jan Haagh
AT&T Global Information Solutions
Wireless Communication Network Division
Utrecht, The Netherlands

Abstract

An interactive software tool has been developed which can be used to assess the coverage of RF transceivers within multifloor buildings. Using AutoCAD as a foundation, this system displays floor plans and permits the user to place transceivers at arbitrary locations within a building. The user can interactively set communication parameters and can view the resulting coverage regions, which depend on the locations and material properties of walls and partitions. Such a tool should prove to be very useful in the specification of wireless LAN systems.

1. Introduction

Wireless LANs are an attractive alternative to wire-based and optical-fiber systems in some environments. A wireless LAN typically incorporates one or more transceivers that are interconnected via a conventional cabled network backbone. These transceivers act as central controllers and communicate at RF or infrared frequencies with data terminal equipment (such as PCs or workstations) that are placed at scattered locations within a building.

A problem in designing a wireless system is that it can be difficult to determine how many transceivers are needed, and where within the building they should be placed. For economic reasons, transceivers should be strategically located so that only the desired areas of the building are covered, and so that neighboring coverage areas do not overlap significantly. Optimum placement of transceivers is complicated by the fact that indoor propagation losses are highly dependent on the type of building, and on the locations and composition of walls within the building.

This paper describes an interactive software system known as the Site Modeling Tool (SMT) which can assist in RF transceiver placement. The SMT, working in conjunction with AutoCAD[1], displays the floor plan for a building and computes estimated coverage areas for transceivers that are placed by the user. These coverage areas are highlighted on the floor plan using color graphics. With this visual information, the user can adjust transceiver locations for improved coverage of the entire building. Transceivers can be added, repositioned or deleted

---

[1] AutoCAD is a trademark of Autodesk, Inc.

interactively, and communication parameters (such as received carrier-to-noise thresholds) can be adjusted by the user as needed.

The SMT supports both single-floor and multifloor buildings. The system recognizes both concrete and cloth partitions, and permits the user to adjust attenuation factors for each. In addition to transceivers, the user can place single-tone interference sources within the building.

The SMT runs in a Windows[2] environment on an IBM-compatible PC. It has been developed as an extension to AutoCAD, which is a popular graphical drawing system. AutoCAD is well suited for the interactive capture and display of building floor plans, and many architects provide floor plans in this form. All SMT software has been written as C-language functions that are accessed through AutoCAD's user interface. It is intended for use by individuals who are not experts at AutoCAD, computing, or wireless LAN design.

In order to enhance portability and future upgrades, the SMT is organized as five relatively independent modules. These are described at a high level in the next section. Sections 3-7 present details of each module separately. Section 8 presents examples which have been generated by the SMT using actual floor plans. Section 9 provides a summary of the paper.

2. Overview of the Site Modeling Tool

The SMT has been developed as five separate modules, as shown in Figure 1. Modularity has been important for simplifying development and for testing SMT software. This approach will permit one system module to be updated at a later time without the need to redesign the rest of the system. For example, if a more sophisticated RF model (used to determine signal attenuation) becomes available in the future, this portion of the system is the only one which needs to be updated.

The Environment Description module contains all available information about the building, including the number of floors, placement of walls, and building materials used. It also maintains information about transceiver locations, interference source locations, and related communication parameters.

The RF Channel module predicts path loss in an indoor environment. When computing path loss for a given point in a building, different equations apply depending on whether when the transceiver of interest is on the same floor or is on a different floor. When on the same floor, two separate (user-selectable) path-loss models have been implemented. One of these is a function of the number of intervening partitions, and the other depends only on distance from the transceiver. When the transceiver is on a different floor from the point of interest, two additional path-loss models are used (and are user-selectable). All of these models depend on the type of building; the currently supported environments include manufacturing plants, hospitals, and office buildings. The SMT can be used for *any* environment for which sufficiently accurate propagation data is available.

The Communication module is built as a hierarchical layer above the RF Channel module. It combines path-loss predictions with additional information to predict the ability to communicate at a given point in the building. For example, a minimum level of received signal strength may

---

[2] Windows is a trademark of Microsoft, Inc.

be specified by the user for this purpose (eg., 10 dB). This module also takes into account the effect of small numbers of single-tone interference sources.

The Site Coverage module has the task of computing the coverage regions of selected transceivers. This involves the iterative calculation of point-to-point communication feasibility, as determined by the Communication module, for many different locations in the building. The output from this module is a contour plot of the coverage area(s).

The last module is the User Interface module. This module is concerned with all entry and display functions of the Site Modeling Tool. It interacts with the user, with AutoCAD, and with the other modules to coordinate all computations.

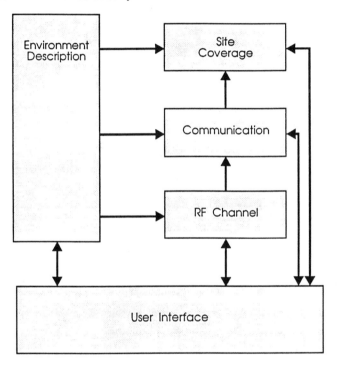

Figure 1. Modules of the Site Modeling Tool. The User Interface provides interactive control of all other modules. The Building Description module contains information that is used by the RF Channel, Communication, and Site Coverage modules.

3.   Environment Description Module

The Environment Description Module essentially serves as a database, containing building information and communications parameters which may be accessed by all other modules of the Site Modeling Tool as needed. Conceptually, this module may be subdivided into 1) a building description component, and 2) a communication description component. For the first of these, useful attributes include construction material (concrete walls, cloth partitions, etc.) for given parts of the building. Other attributes that are stored here include partition attenuation factors, and the distance between floors if the building comprises more than a single story.

The communication description component maintains the locations of transceivers and single-tone interference sources. Other attributes that are stored here include transmitting power, minimum carrier-to-noise and carrier-to-interference ratios, receiver sensitivity, and interferer frequencies. After coverage contours have been computed for a specific transceiver location, they are stored here as well.

4. RF Channel Module

4.1. Overview

The SMT incorporates four RF channel models developed by Seidel and Rappaport [Seid92]. These models utilize relatively simple prediction rules to estimate the wideband signal loss between any two points inside a building. Two of the models are used with the transmitter and receiver are on the same floor, and the remaining two models apply when different floors are involved. In each case, the user may select which model is used. The terms that are used in the models are summarized in Table 1.

Table 1: Summary of the variables used to predict path loss.

Variable	Description	Units
d	distance separating transmit and receive points	m
d_0	reference distance	m
PL	path loss	dB
$\overline{PL}$	mean path loss	dB
n	mean path loss exponent	—
λ	carrier wavelength	m
p	number of soft partitions in path	—
q	number of concrete partitions in path	—
AF_p	attenuation factor per soft partition	dB
AF_q	attenuation factor per concrete partition	dB
FAF	floor attenuation factor	dB

All models in the Site Modeling Tool assume that the mean path loss $\overline{PL}$ is an exponential function of the distance d to the power n from the transmitter. Since loss is calculated in dB, n is multiplicative. In microcellular and indoor environments, a reference distance $d_0 = 1$ m is typically used. For the SMT, all signal levels are referenced to this distance as follows:

$$PL(d_0) = 20 \times \log_{10}\left(\frac{4\pi d_0}{\lambda}\right)$$

4.2. Single-floor models

Distance-dependent path loss model: This model does not depend explicitly on the partition locations in the floor plan. Instead, different values for n are used, depending on the type of building. It may be expressed as

$$\overline{PL}(d) = PL(d_0) + 10 \times n(samefloor) \times \log_{10}(d / d_0).$$

In free space, $n = 2$. For the SMT, the user selects the appropriate building environment, and the program assumes a value for n that has been empirically determined. Several suggested values are shown in Table 2. The user may also enter a value for n directly.

Table 2. Empirically determined values for n.

Type of building	n
Office building	3.3
Hospital	3.0
Grocery store	1.8
Warehouse	1.6
Airport	2.0
Manufacturing plant	1.8
Retail store	2.5

Partition-attenuation-factor path loss model: This model assumes that the signal attenuates as in free space ($n = 2$), and attributes all additional path loss directly to intervening soft partitions and concrete walls:

$$\overline{PL}(d) = PL(d_0) + 10 \times n(freespace) \times \log_{10}(d / d_0) + p \times AF_p + q \times AF_q.$$

Typical values for AF_p and AF_q are 1.4 dB/partition and 2.4 dB/partition, respectively.

4.3. Multifloor models

Distance-dependent path loss model: Both multifloor models are used only when the transmit and receive locations are on different floors of the building. This model is similar to the first model in Section 4.2, except that a larger value of n is chosen to account for the increased attenuation of intervening floors. This may be represented as follows:

$$\overline{PL}(d) = PL(d_0) + 10 \times n(multifloor) \times \log_{10}(d / d_0).$$

Floor-attenuation-factor path loss model: This model is also similar to the first model given in 4.2. However, the additional path loss caused by intervening floors is now added explicitly as the quantity FAF:

$$\overline{PL}(d) = PL(d_0) + 10 \times n(samefloor) \times \log_{10}(d / d_0) + FAF.$$

Typical values for FAF lie in the range 13 to 30 dB and have been estimated empirically. The FAF is a function of the number of intervening floors, and is stored as a look-up table within the SMT.

5.   Communication Module

This module predicts communication feasibility for a given point within a building. It relies on estimates of path attenuation from the RF Channel module, and on information provided by the Environment Description module.

In order to determine whether communication is feasible, this module compares the received carrier power and carrier-to-noise ratio (CNR) with the receiver sensitivity and minimum required CNR at the point under consideration.   Further, if a small number of single-tone interference sources are present, the received carrier-to-interference ratio (CIR) is compared with the minimum allowed CIR.   (The receiver sensitivity is a function of each interferer's transmitting frequency.) Communication is deemed feasible if the received signal levels are all above the minimum user-specified thresholds.

The calculation of noise power (N), in addition to thermal noise, also incorporates the effect of environmental (or ambient) noise. Recent studies by the Mobile and Portable Radio Research Group at Virginia Tech indicate that the level of this noise is typically about 18 dB above the thermal noise floor [Blac93].

6.   Site Coverage Module

This module predicts areas of coverage within a building and displays these areas directly on the floor plan.   The coverage area for a transceiver is the set of nearby points for which communication is feasible, as determined by the Communication module.

By default, the SMT assumes that signals propagate equally in all directions from a transmitter. The default coverage area is therefore circular for a given floor (or spherical if three dimensions are considered).   When interference sources are present, or when walls are present and the partition-dependent path loss model is used (Section 4.2), the default circular shape is perturbed according to the associated attenuation factors and CIR thresholds. The SMT further assumes that coverage areas are always simple closed regions — no holes are present in a coverage area, and a single coverage area is never subdivided into separate, noncontiguous regions.   These appear to be safe assumptions except for degenerate cases when interference sources are in close proximity to a transceiver.

With these assumptions, it is sufficient for the Site Coverage module to detect the points near the *border* of the coverage area. Because of the assumption of simple closed regions, it is not necessary to confirm that communication is feasible at every internal point of the coverage area. Since the exact shape of the coverage area is not known in general, a search must be performed to locate border points. For graphical highlighting on the floor plan, these points may be used as vertices of a polygon which approximates the border.

The Site Coverage module searches for border points along lines that radiate from the transceiver's location. The search lines are uniformly spaced. Normally 36 border points are found, although the number is user-selectable. It is computationally prohibitive to increase the number of points substantially.

The first border point is found horizontally and to the right of the transceiver. Without loss of generality, assume that the transceiver is located at the origin of a two-dimensional (2D) Cartesian coordinate system. An initial point is first chosen at the reference distance d_0 along the x axis. Points at intervals along the x axis are evaluated until communication is no longer feasible. At this time, the first contour point is assumed to lie at the midpoint of the two points most recently evaluated.

For subsequent points, the search occurs along an adjacent radial line, beginning at the same distance from the transceiver as the most recently detected border point. The search proceeds toward or away from the transceiver, depending on the communication feasibility of this point. The search process ends after all directions for the given number of border points have been searched.

7.   User Interface Module

The purpose of the user interface is to provide a simple interactive mechanism for controlling the operation of the SMT. It allows the user to set building and communication parameters, place transceivers and interference sources, move from floor to floor, and direct the calculation of coverage areas. This module provides textual and graphical feedback to the user, and directs the operation of all other SMT modules.

As described in Section 1, the SMT has been developed as an extension to AutoCAD. The user interface therefore relies heavily on AutoCAD to provide an operating environment. A mouse and keyboard are used as input devices, and feedback is provided on the monitor of the PC.

Some of the more prominent provisions of this module are as follows:

- Because AutoCAD serves as the foundation of the system, full AutoCAD functionality is available to the user. The user can therefore enter floor plans from scratch, or import them from files.

- The user can select an "environment type," such as office building or warehouse, as described in Section 4.2.

- The user can interactively add, reposition, and delete any number of transceivers and interference sources.

- The user can interactively select communication parameters such as transmitter power and minimum CIR.

- The user can initiate the calculation of coverage areas for transceivers. These areas are highlighted graphically and in color on the floor plan.

8.   Examples of operation

To illustrate SMT operation, consider the floor plan that appears in Figure 2. This is one floor of a building on the Virginia Tech campus, and contains several offices, classrooms, and storerooms There are two elevator shafts and stairwells. This drawing was provided to us in AutoCAD form. Overall dimensions of the building as shown are 57 m by 91 m.

Instead of using this floor plan immediately, it is best if extraneous details are removed from consideration. Examples of these are arcs representing doors, lines representing stairs, and outside walkways. If the architect has placed such details on separate AutoCAD *layers*, it is a simple matter to ignore them during path-loss computation. If they are not on separate layers, the user should remove them (or transfer them to new layers) for reasons of accuracy and efficiency.

Figure 3 shows the same floor plan as before, but after the unnecessary details have been removed. In the figure, the user has initiated the SMT and has placed a transceiver near the center of the building. The resulting coverage for this transceiver is shown as two bold contours with a dotted fill pattern. The inner and outer contours represent "safe" and "marginal" coverage areas, respectively, and are separated by a 5 dB margin. The partition attenuation factor path loss model has been used (Section 4.2), and 36 points have been located on each contour.

Figure 4 illustrates the effect of moving the transceiver by a small amount. By simply translating the transceiver to the corner of the original room, the shape of the coverage contour is altered significantly. Figure 5 demonstrates the effect of a single-tone interferer. This is the diamond shape located diagonally across the room containing the transceiver. The transceiver is in the same location as for Figure 4. Notice that the coverage area opposite the interferer is unchanged, but the coverage near the interferer is altered dramatically.

Now assume that the user moves to the floor immediately above the one under consideration. The user places a transceiver on that floor, and returns to the floor being discussed. The coverage for the second transceiver can then be displayed, as shown in Figure 6. This illustrates that the multifloor path loss models do not consider partitions individually, as illustrated by the near-circular shapes of the two new contours. The interferer has affected the marginal coverage area.

Finally, a floor plan for a different building appears in Figure 7. Three transceivers have been placed, and their coverage areas highlighted. After viewing these coverage areas, a system designer would adjust their locations and consider adding a fourth transceiver to obtain complete coverage.

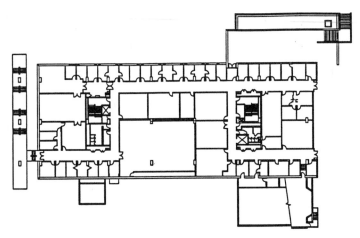

Figure 2. Example floor plan. This is one floor of Whittemore Hall on the Virginia Tech campus. This drawing was generated directly from an AutoCAD file provided by the university's Facilities Planning and Construction department.

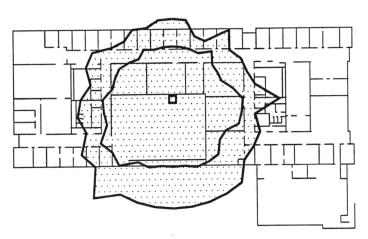

Figure 3. Coverage contours for one transceiver. Unnecessary details have been removed from the drawing in the previous figure, and concrete partitions only are assumed. Notice that the coverage area is reduced near regions where many partitions are present. The drawing is more appealing when viewed in color on the PC's monitor.

194

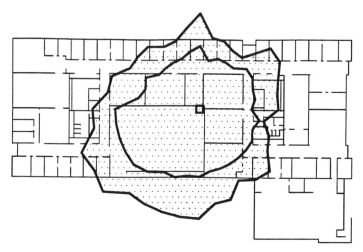

Figure 4. Coverage contours for repositioned transceiver. The transceiver of the previous figure has been moved to the right by a small amount.

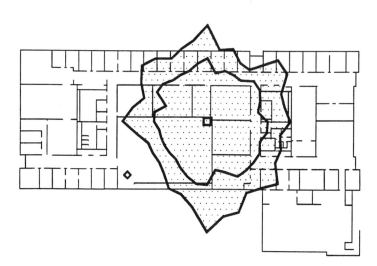

Figure 5. Effect of single-tone interferer. The transceiver has not moved from its location in Figure 4. The coverage area is changed considerably by the interference source.

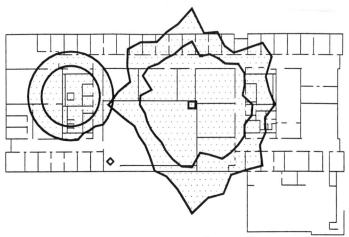

Figure 6. Addition of coverage contours for transceiver on different floor. Transceivers not on the floor being viewed are distinguished by squares with thinner lines. The coverage area is nominally circular, but is perturbed here by the interference source.

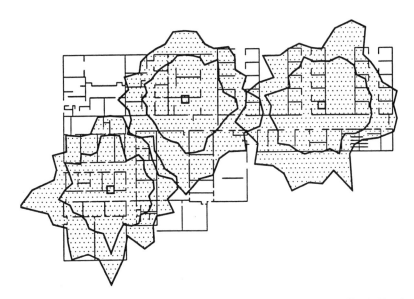

Figure 7. Second example floor plan. This is one floor of Virginia Tech's Veterinary Medicine building. Three transceivers have been placed as a first attempt to provide complete floor coverage.

9. Summary

This paper has described a new visualization tool that can be used to predict coverage areas for wireless LANs within buildings. The system permits the user to place transceivers and interference sources interactively within a floor plan that has been generated using AutoCAD. After placement, the user can direct the system to compute coverage areas which are highlighted directly on the floor plan. Coverage areas are calculated as a function of transceiver parameters, locations of walls and soft partitions, and building type. For versatility, most parameters are user-selectable. The system supports both single-floor and multifloor buildings.

In order to make the Site Modeling Tool interactive, relatively simple propagation models have been implemented. These models do not require computationally intensive techniques such as ray tracing. Instead, they depend only on distance and the number of intervening building partitions. The models were developed using extensive empirical measurements in earlier work [Seid92], and provide useful results in many environments.

The Site Modeling Tool satisfies an important need for system designers of wireless LANs. The tool is simple to use, it accommodates many different types of buildings, and it provides visual feedback that is intuitive and easy to understand.

10. References

[Seid92] S. Y. Seidel and T. S. Rappaport, "914 MHz Path Loss Prediction Models for Indoor Wireless Communications in Multifloored Buildings," *IEEE Transactions on Antennas and Propagation,* vol. 40, no. 2, pp. 207-217, Feb. 1992.

[Blac93] K. L. Blackard and T. S. Rappaport, "Impulsive Noise in Personal Communication Systems," *IEEE Journal on Selected Areas in Communications*, vol. 11, no. 7, pp. 991-1000, Sept. 1993.

15

Influence of Slow Frequency Hopping and Antenna Diversity Techniques in European GSM System

Fabrice de SEZE

ALCATEL RADIOTELEPHONE

32, avenue Kléber

92707 Colombes Cedex France

Abstract—Antenna Diversity (AD) and Slow Frequency Hopping (SFH) are used in GSM cellular system to improve significantly the transmission quality. The former technique combats multipath fading and shadowing while the latter additionally averages the effects of co-channel interference. The aim of this paper is to identify the optimum use of SFH and AD techniques in urban environment with cells of radius less than 1 km and with mobiles moving at a speed of 50 km/h and below.

In the first part, the simulation tool is described and optimum parameters for each method (choice of frequency in SFH, correlation of receiving signals in AD etc.) are proposed. Then improvements introduced by these different techniques are compared. The simulation results presented assess the efficiency domains of each of these techniques depending on the mobile velocity. Finally, assessment of the performance achieved when both techniques are used simultaneously is presented.

The simulation results show that in urban environment both techniques allow to have a system working with a lower co-channel interference level. However SFH is more efficient than AD for mobiles whose velocity is below 20 km/h. Moreover C/I versus BER curves reveal that the best performance is obtained with a system using SFH and AD simultaneously and that improvements obtained by each technique are additive.

1. Introduction

In order to combat multipath fading, shadowing and the effects of co-channel interference, European GSM cellular systems have implemented Antenna Diversity (AD) and Slow Frequency Hopping (SFH). In this paper, we compare these different techniques in urban environment.

We begin with a presentation of the simulation tool and a description of the simulation parameters. Then a first set of simulation is performed, considering fast fading only. The model is improved and other aspects of cellular system in urban environment—shadowing, mobility and traffic— are considered. The simulation results show that the benefit of SFH increases when these aspects are incorporated in the model. For mobiles moving at a speed below 20 km/h, the performance gain is better with SFH than with AD while the best performance is obtained when both techniques are combined.

198

2. Implementation

2.1. Antenna Diversity

Between the different types of space diversity, we can identify micro diversity (received signals from several diversity antennas installed at the receiver location are combined) and macro diversity (received signals received from different base stations are combined).

Although, in GSM, the use of AD is not mandatory, ALCATEL proposes this technique in its base stations as an option. The signals received from two antennas are combined at the input of the channel decoder (soft bits from the two branches are added). Figures 2 shows how antenna diversity is implemented in ALCATEL base stations.

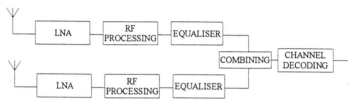

Figure 1: Antenna Diversity Implementation in ALCATEL Base Station.

The improvement introduced by such a system will depend on the correlation of the two received signals. This correlation varies according to parameters like the environment, the antenna characteristics, the antenna arrangement etc.

2.2. Slow Frequency Hopping

Frequency hopping is a proven technique to eliminate the disturbing effects due to both multipath fading and co-channel interference and has been already presented in detail in [1] and [2]. In GSM, on each carrier, the TDMA frame is divided in 8 time slots. Each mobile transmits (or receives) a burst on a fixed frequency during one time slot and then must hop before the time slot on the next TDMA frame. Efficiency of SFH depends on the difference between the various frequencies among which the mobile hops on. This difference must be higher than the coherence bandwidth. Multipath fading is far more disturbing with a slow moving mobile. The difference can be observed with mobiles moving at different speed (figure 5 to 8). In GSM, SFH is mandatory in mobile stations and optional in base stations. SFH algorithm is cyclic or pseudo random and is fully described in [3]. The frequency on which the mobile transmits (or receives) depends on the Hopping Sequence Number (HSN) and on the Mobile Allocation Index Offset (MAIO).

3. Simulation tool

The simulation tool is based on the GSM algorithms for full-rate speech[3 to 6]. The 260 bit speech frame corresponding to 20ms is coded and interleaved. Then it is carried over 8 bursts. The modulation is GMSK (BT=0.3) and we equalise the bursts with a 16 state Viterbi equaliser. The channel simulator program has been developed in RACE 1043 project[7]. It can simulates different channel types. The following simulations will use a COST 207 TU channels[6] (6 Rayleigh fading taps channels). Statistics are performed over 5000 burst (i.e. 1250 speech frames, 25 s). Frame Erasure Rate (FER) is the selected performance for evaluating the radio link. A received speech is erased if one of the 50 most important bits of a speech frame is erroneous. Interfering signals are not represented in the block diagram. They are affected by uncorrelated radio channels (6 Rayleigh fading taps channels see [6]).

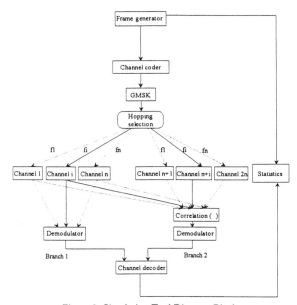

Figure 2: Simulation Tool Diagram Block.

In a narrow band system, characteristics of the radio channel can be modelled by both a fading process describing the channel and an additive white Gaussian noise describing the noise in the receiver. Because of the way antenna diversity is implemented, only the fading processes need to be correlated in the simulation tool.

Fading process is simulated by generating a distributed complex random variable filtered with a Doppler spectrum. If x and y are two uncorrelated variables, a third variable $z = \rho x + y\sqrt{1-\rho^2}$ can be generated.

Then x and z have a correlation factor of ρ according to the definition of two complex random processes [8]:

$$\rho_{xz} = \frac{E\{[x - E\{x\}][z^* - E\{z^*\}]\}}{\sqrt{E\{|x - E\{x\}|^2\}E\{|z - E\{z\}|^2\}}}$$

Due to the linearity of the filtering process, the correlated signals can be generated by correlating the two complex variables before filtering or by correlating the two received signals. The latter solution has been chosen for the following simulations.

A fading process is associated to each frequency (channel 1 to channel n and for the diversity branch, channel n+1 to channel 2n in figure 2). The hopping algorithm is pseudo random.

It is very important to note that the tool simulates the channel over a short period of time. That is why only fast fading is considered in the first part of simulations. This means that the mean power level of the main signal and of the interferers are constant during the 25 seconds simulated. So, a set of simulations is performed with different mean power values of the interferers. This set of short-term results is used as a look-up table to include long terms effects of shadowing, mobility and traffic in the evaluation of the system performance.

A first set of simulations allows to choose the correlation factor and the number of frequencies for hopping. Figure 3 shows that the improvement of the correlation factor of the signals coming from the two different antennas introduce an important gain between 0.5 and 1. On the contrary, it is not very useful to reduce the correlation of the two signals for a correlation factor under 0.5. Moreover, figure 4 shows that up to 4 frequencies, we improve the significantly the link budget, while the gain is far below over 4 frequencies. Therefore we have chosen correlation factors of 0.5 for AD and 4 frequencies for SFH in all the following simulations.

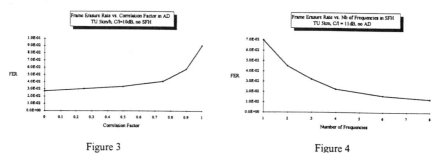

Figure 3   Figure 4

4. Performance Evaluation

4.1. Fast Fading

AD and SFH as well protects against fast fading. We performed a set of simulations with one interferer in an urban environment with mobiles moving at various speeds. When SFH is simulated, the interfering signal has the same hopping sequence as the wanted signal (following the GSM recommendation [6]). As a consequence, only improvements due to frequency diversity (which mainly eliminates multipath fading) is considered as this point.

We depict in figure 5 to 8 that for faster moving mobiles, the efficiency of SFH decreases. A first conclusion can be drawn : **as far as fast fading is concerned**, AD improves the system in urban environments and SFH is as efficient as antenna diversity for slow moving mobiles. Moreover, SFH and AD improvements are cumulative.

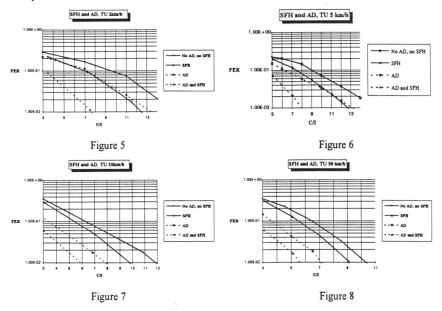

Figure 5

Figure 6

Figure 7

Figure 8

4.2. Shadowing and Mobility

It is out of the scope of this chapter to give a full system performance evaluation. But we would like to highlight some advantages of frequency hopping when shadowing and terminal mobility are considered. We encourage the reader to refer to [9] and [10] and to consider this chapter as a complement of these papers.

If we consider shadowing and terminal mobility, we cannot assume a constant mean C/I for a given reuse distance. For simplicity, we consider only one interfering cell. Moreover this interfering cell will use the same frequencies as the main cell but with a different hopping sequence number(HSN). As a consequence, only one mobile will interferer at a given time but this mobile will change from one burst to another burst. Therefore, as we can see in figure 9, the system will benefit from an interferer diversity when the mobile will transmit. This is not the case (figure 10) for the down link because we assume no power control in the base station and consequently the power level of the interferer is the same at each burst.

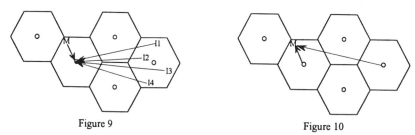

Figure 9   Figure 10

A set of simulation has been performed using the simulation tool. But in comparison to the simulations described in 4.1., 4 interfering signals are considered, each using the same HSN (which is different from the HSN of the main signal), using a different MAIO and finally each one having a different mean power level. As a consequence, at every burst transmission, the main signal will have one and only one interferer but this interferer is different from one burst transmission to another one.

Figure 11 shows the FER against the mean C/I where : $\dfrac{\overline{C}}{I}(dB) = 10\log_{10}\dfrac{4\times10^{\frac{C_{(dB)}}{10}}}{\sum\limits_{i=1}^{i=4}10^{\frac{I_{i(dB)}}{10}}}$.

We can see that for a given $\dfrac{\overline{C}}{I}$, the performance in terms of FER can vary significantly, the worst case being $I_1=I_2=I_3=I_4$ which corresponds to the line. In order to simplify the simulations, we will consider in the rest of this paper, that the FER for a given $\dfrac{\overline{C}}{I}$ will be this worst case value.

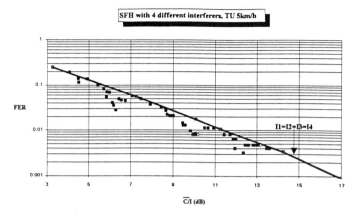

Figure 11

Figures 12 shows the cumulative distribution function of FER with AD and SFH in a TU2 environment for the uplink. The path loss (in dB) is given by $P_r = P_{r_0} - 10\gamma \log_{10} \frac{r}{r_0}$ where the path loss slope γ=3.5. Shadowing is modelled by a log-normal distribution of the local mean. The standard deviation of the log-normal probability density function (pdf) is 6dB for the wanted and for the interfering signals. The mobiles are uniformly distributed in the 1 km radius cells. The distance between the main cell and the interfering cell is 3.5 km.

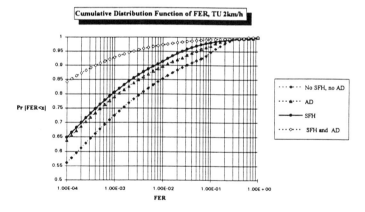

Figure 12

From a system performance point of view, it will be required to have a good quality 95% of the time. In table 1, we consider that good quality means a FER below 3% and we compare in different configurations what should be the minimum distance between the main cell and the interfering cell (cell radius=1).

	TU 2km/h	TU 5km/h	TU 20km/h	TU 50km/h
No SFH, no AD	4.4	4.3	3.6	3.4
AD	3.8	3.8	3.1	2.8
SFH	3.5	3.5	3.2	3.1
AD and SFH	2.8	2.6	2.4	2.3

Table 1: Reuse distance

4.3. Traffic

Up to now, we have assumed that each slot of the interfering cell was occupied. Of course, it is not the case in a real system. In this chapter, we show that the benefits of SFH will increase when an Erlang-B traffic distribution model is incorporated. Even though we continue to consider only the uplink, the downlink will benefit from the same advantages.

We consider that a cell with 4 carriers (one per frequency) has been dimensioned to handle 20.2 Erlangs (blocking probability = 2% if we assume that over the 32 available slots—in GSM a carrier contains 8 slots—, 4 are allocated to signalling). Then the probability to have N occupied slots over the 28 traffic slots

is
$$
\begin{cases}
P(N) = \dfrac{A^N}{N!} P(0) \\
P(0) = \dfrac{1}{\displaystyle\sum_{i=0}^{i=N} \dfrac{A^i}{i!}}
\end{cases}
\qquad \text{with } A=20.2 \text{ and } 0 \leq N \leq 28.
$$

Without SFH, the probability to have a mobile interfering is $p(1) = \displaystyle\sum_{i=1}^{i=28} \dfrac{i}{28} P(i)$

With SFH, each interfering mobile interferers one fourth of the time. The probability to have $k (0 \leq k \leq 4)$ interfering mobiles is $p(k) = \displaystyle\sum_{i=k}^{i=28-4+k} P(i) \dfrac{\binom{28-i}{4-k}\binom{i}{k}}{\binom{n}{4}}$

Figure 13 shows the cumulative distribution function obtained with the traffic model in TU 2km/h in the condition defined in 4.2.

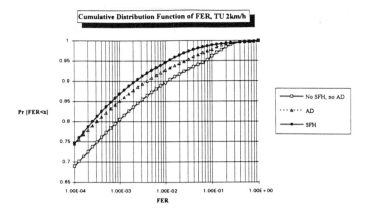

Figure 13

Table 2 shows the reuse distance (as defined in 4.2) when traffic is considered. As we can see in this table, the gain in terms of system capacity is the same with AD and with SFH when the mobile moves at 20km/h in a urban environment.

	TU 2km/h	TU 20km/h
No SFH, no AD	3.8	3.4
AD	3.5	2.9
SFH	3.1	2.9

Table 2: Reuse distance

5. Conclusion

In this paper, we have compared AD and SFH and we have shown the benefits of SFH with slowly moving mobiles. Performances have been evaluated with only one interfering cell. Future work should be done with more than one interfering cell. In any case, the expected results should confirm that SFH improves considerably the system capacity in urban environments.

Even though we focused on capacity aspects, this paper has also demonstrated that for a given capacity, introduction of SFH and AD improve significantly the transmission quality which is today the main request from the GSM operators.

206

6. Acknowledgements

The author expressed his gratitude to Dr. V. Kumar, Dr C. Evci and Mr. M. Delprat for very useful comments.

7. References

[1] D. Verhulst, M. Mouly and J. Szpirglas, Slow Frequency Hopping Multiple Access for Digital Cellular Radiotelephone, IEEE JSAC-2, July 1984

[2] J. Dornstetter and D. Verhulst, Cellular Efficiency with Slow Frequency Hopping , IEEE JSAC-5, June 1987

[3] Recommendation GSM 05.02, Multiplexing and Multiple Access on the Radio Path, v 3.4.1, January 1990

[4] Recommendation GSM 05.03, Channel Coding, v 3.5.1, January 1990

[5] Recommendation GSM 05.04, Modulation, v 3.1.1, January 1990

[6] Recommendation GSM 05.05, Radio Transmission and Reception, v 3.11.0, January 1990

[7] C. Evci, Pan-European Project for Third Generation Wireless Communications, Winlab Workshop, April 1992

[8] A. Papoulis, A. : Probability, Random Variables and Stochastic Processes. McGraw-Hill. 1965

[9] R. Wyrwas and J.C. Campbell, Radio Topology Design with Slow Frequency Hopping for Interference Limited Digital Cellular Systems, Vehicular Technology Conference 1991

[10]   A. Safak and R. Prasad, Effects of Correlated Shadowing Signals on Channel Reuse in Mobile Radio Systems, Trans. on Vehicular Technology November 1991

16

Network Architecture and Radio Link Performance of MOBITEX® Systems

Reza Alavi
Manager, Network Analysis
RAM Mobile Data
10 Woodbridge Center Drive
Woodbridge, NJ 07095

M. Mobeen Khan
Network System Engineer
RAM Mobile Data
10 Woodbridge Center Drive
Woodbridge, NJ 07095

ABSTRACT

MOBITEX is a worldwide standard for wireless wide area packet-switched data communications. Because all MOBITEX systems provide documented open interfaces, any company can enter the mobile data business by supplying mobile and portable radio modems and application software to users of a MOBITEX system. This paper provides an overview of the MOBITEX system including the MOBITEX network architecture, network protocol, public and private data network connectivity, RF system design, and the radio link protocol, ROSI. Several MOBITEX system features, such as Host Group Addressing (HGA) and store-and-forward, are also described. Finally, simulation results for throughput and response times of the ROSI protocol are presented.*

1.0 Introduction

MOBITEX was developed by Ericsson and Swedish Telecom and was first placed into commercial operation in Sweden in 1986. MOBITEX networks are operating in North America, many countries in western Europe, and Australia. MOBITEX systems are optimized for the specific demands of data communications. They provide fast response times, high spectral efficiency, and features such as real-time and store-and-forward messaging, closed user groups, dedicated system connections, broadcast and multi-address messaging, and host group addressing. Worldwide standardization of all MOBITEX systems and the compatibility of future development of advanced features and functions are ensured by the MOBITEX Operators Association (MOA). MOA, which also manages the MOBITEX interface specification [1], is an independent organization comprising all MOBITEX system operators.

RAM Mobile Data[1] provides nationwide mobile data service in the U.S. using MOBITEX. RAM's system links more than 6300 cities and towns, reaching more than 90% of the urban business population.

---

[1] RAM Mobile Data USA Limited Partnership is a business venture between RAM Broadcasting and BellSouth

2.0 Network Architecture

MOBITEX systems are hierarchical in structure and may contain up to six levels of network nodes. At the top of the hierarchy is the Network Control Center (NCC) which is responsible for all network management. The NCC is equipped with a VAX computer, printers, and terminals to enable system personnel to perform customer activation, billing, and network management functions. At the bottom, multi-channel base stations provide wireless access to mobile users. There can be up to four intermediate levels of nodes between the NCC and base stations.

Below the NCC is the main exchange which switches messages between lower level nodes and also interacts with the NCC. Below the main exchange, regional and local switches are deployed depending on the traffic volumes in the areas these switches serve (see Figure 1). In RAM's system, these areas are called LATAs (Local Access Transport Areas). LATAs are interconnected through long distance service providers using either the HDLC or X.25 protocol running at 9.6 or 56 kbps.

Host computers access RAM's MOBITEX system either directly through a local switch port or through a Front-End Processor (FEP) connected to a local switch (using an X.25 connection). The data rate of the host connection can be either 9.6 or 56 kbps. The FEP supports both host interfaces and conversions for protocols such as SNA, TCP/IP, and LU2. All land-line connections between MOBITEX nodes and customer sites are made through data service units (DSUs) that are monitored at the NCC using an AT&T Paradyne COMSPHERE 6800 Network Management System.

Base radio stations are the lowest nodes in the MOBITEX hierarchical structure and they are the connecting point for portable and mobile radio modems. Base stations switch traffic between users communicating through the same base station.

If the connection to a higher node is lost, all MOBITEX nodes are capable of running autonomously. Traffic between users connected to subordinate nodes is not affected. The RAM MOBITEX system has an overall system availability of greater than 99.9 percent. Since its deployment, RAM's MOBITEX system has proven extremely reliable in the wake of several natural disasters and emergency situations nationwide.

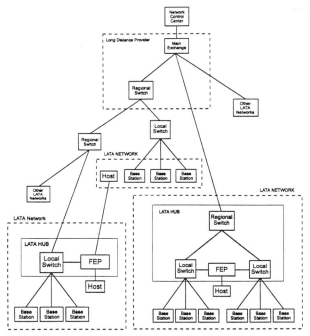

Figure 1. MOBITEX Network Architecture

3.0 Network Communications

To facilitate interconnections between network components, MOBITEX systems adhere to the OSI (Open Systems Interconnection) seven layer model. As data passes through the system, each layer adds its unique protocol information. Network layer communications in MOBITEX is accomplished using packets called MPAKs. An MPAK contains from one to 512 bytes of user information. Messages are sent in ASCII text format or as data coded in any application-dependent format. Higher Protocol Data (HP Data) is used for messages longer than 512 bytes and allows the use of higher level protocols. Figure 2 shows the OSI representation of connections at the network, data link, and physical layers in MOBITEX.

All subscribers are assigned a unique MOBITEX Access Number (MAN). Group broadcast and distribution list addressing allow messages to be sent to several MANs simultaneously. A Closed User Group (CUG) provides subscribers with a virtual private network where only intra-CUG communication is permitted. The MOBITEX system also allows multiple host connections using one MAN to provide added reliability and load balancing. The network will route user messages to the closest available host. This is referred to as Host Group Addressing (HGA).

210

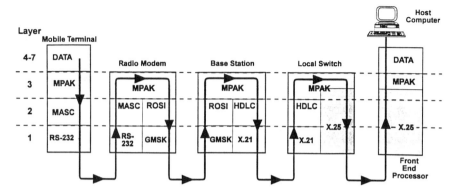

Figure 2. Mobile-to-Host Communication Protocols

Packet routing is performed at the lowest possible level in the hierarchy using roaming information stored in each node. This information is updated in the relevant nodes each time a mobile roams to a new base station. A store-and-forward capability is available for subscribers who are temporarily out of contact with the system when messages are sent to them. Seamless nationwide roaming guarantees the delivery of any stored messages regardless of the geographical position of the mobile when it re-establishes contact with the network.

4.0 Radio Link Communications

Base radio stations in a MOBITEX system are deployed in a cellular-like pattern, as shown in Figure 3. The pattern employs the same engineering principles used in cellular telephony relative to interference and frequency management, and cell splitting. MOBITEX cell coverage design adheres to the 90/90 principle (i.e., radio coverage extends to at least 90 percent of a designated area at least 90 percent of the time). Unlike cellular telephone systems, the MOBITEX radio modem evaluates the signal strengths of the current and neighboring base stations and roams only if certain criteria are satisfied. The major roaming criteria are:

1) The received signal strength (RSSI) from a neighboring base station is greater than the current base station's RSSI by a significant amount

2) The RSSI of the current base station is too low

3) The radio modem has tried unsuccessfully to access a base station or retransmit a message

4) Base station contact is lost.

When evaluating RSSI, a radio modem continuously monitors the current base and updates the neighboring base stations' RSSI values once every 10 seconds. These measurements are averaged with previous values to reduce the adverse effects of log-normal fading and shadowing. When a mobile terminal roams to a new

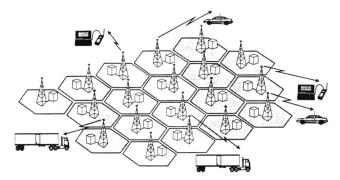

Figure 3. MOBITEX Base Station Deployment

base station, it transmits a registration packet to the new base station to inform the system that all subsequent traffic to this mobile terminal should be routed through the new base station.

Because data transferred between a mobile and base station must travel as radio waves, environmental conditions such as interference, noise, fading, and signal loss due to shadowing can introduce errors into that data stream. MOBITEX uses the ROSI radio link protocol to ensure a reliable and efficient transmission path between the mobile terminals and the base stations. There are three methods incorporated in the ROSI protocol to ensure that the probability of successful transmission through the radio link approaches 100 percent. The three error detection/correction methods used in the ROSI protocol are:

1) Block coding -- Data is grouped into blocks of 18 bytes. A CCITT standard 16-bit (2-byte) CRC is appended to each block to detect all one and two-bit errors, and all bursts errors up to 16 bits in length. Each byte of the 20 x 8 block is further coded using the [12,8] Hamming code to correct single bit errors.

2) Interleaving -- After block coding, the resulting 20 x 12 block is interleaved before transmission to protect against burst errors due to the effects of fading. After interleaving, bursts of up to 20 errors can be corrected.

3) Automatic Repeat Request (ARQ) -- When an error is detected in the radio frame but correction has not succeeded, selective repeat ARQ is applied to enable the sender to retransmit only those blocks with errors.

After interleaving, a scrambling sequence is used to generate a pseudo-random pattern of 1s and 0s. Scrambling is not a form of error correction but is used to ensure that the radio receiver remains locked to the center of the frequency band (long bursts of continuous 1s or 0s may cause the radio to drift).

Finally, before transmission, a 56-bit frame head is added which includes information used to identify the MOBITEX system and base station.

Channel access for passage of traffic between mobile terminals and a base station is controlled by the base station. The basic channel access algorithm uses a reservation, slotted ALOHA protocol tuned to the characteristics of the data traffic. In this scheme, the base station sends a "free" (FRI) signal that indicates a number of available random slots. Each mobile terminal picks a random slot and sends a short data packet or access request to the base station. At the end of the random slots, the base station acknowledges the short packets and grants permission to each mobile terminal to transmit its longer message until all request have been satisfied. The base station then issues another FRI signal. The base station can, however, transmit downlink traffic any time between FRI signals since it operates in full duplex. Additional channels can be allocated by each base station to optimize total radio link throughput under various traffic conditions. Both trunking and fleet division can be supported.

Another important feature of ROSI is the battery saving protocol. This protocol effectively increases the battery life of a portable radio modem. Radio modems operating in battery saving mode stay in a low power standby mode and "wake up" (switch to the operating state) periodically to listen for base station signals. These signals include information on timing and roaming, and contain a list of downlink messages. A mobile terminal with a message pending will remain in the operating state until the message is successfully received and acknowledged.

5.0 MOBITEX Radio Link Performance

Performance of the radio link can be measured in terms of throughput and response time and is dependent on the following criteria:
- Message length distribution
- Radio link data rate
- Overall traffic load.

Uplink and downlink messages have exponential and uniform length distributions, respectively. Experience has shown that the mean message length of uplink traffic is seven blocks, whereas the mean message length of downlink traffic is 16 blocks. The radio link data rate is 8 kbps. Using these parameters, the performance of the MOBITEX radio link is shown in Figures 4 and 5 for different downlink traffic rates. The curves indicate that uplink throughput increases when the downlink traffic rate decreases, since a decrease in the downlink traffic rate enables the base station to issue more FRI signals. Also, under heavy load conditions, the radio link throughput shows only gradual degradation.

Successful transmitted packets per hour x1000

○ 0.5 Packets/second downlink   ◊ 1.0 Packets/second downlink   □ 1.5 Packets/second downlink

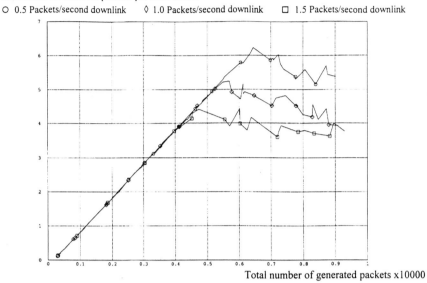

Total number of generated packets x10000

Figure 4. MOBITEX Radio Link Throughput

Average delay per packet in seconds

○ 0.5 Packets/second downlink   ◊ 1.0 Packets/second downlink   □ 1.5 Packets/second downlink

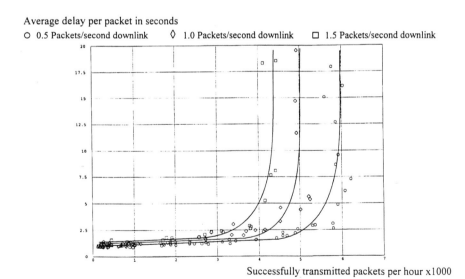

Successfully transmitted packets per hour x1000

Figure 5. MOBITEX Radio Link Delay Response

6.0 Competitive Position

In the wireless communications industry, enhanced paging services (Skytel, EMBARC, Mtel's NWN, etc.), SMR systems (ARDIS, Racotek, etc.), ESMRs (NexTel), and CDPD systems are, or will be, RAM's competitors. The many unique capabilities of MOBITEX (seamless nationwide roaming, international service, store-and-forward messaging, battery saving protocol, closed user groups, and flexible data protocols) are major characteristics that give RAM Mobile Data a distinct advantage. RAM's MOBITEX system can fulfill the wireless messaging needs of most businesses today.

MOBITEX is a stable system based on state-of-the-art technology. RAM Mobile Data, Ericsson, and Eritel continue to upgrade MOBITEX technology in order to keep it in the forefront of wireless packet switching systems. The newest revision (R14) provides increased reliability, availability, security, host group addressing, alternate network pathways between nodes, multiple connection between nodes, on-line network reconfiguration, positive acknowledgment, ESN (Electronic Serial Number) validation, and several enhanced network management features. RAM continues to build strategic alliances with application, platform, and terminal manufactures to provide MOBITEX-enabled software and hardware.

Future enhancements to MOBITEX include increased data rates between network nodes and in the air link protocol, improved distributed switching using regional NCCs, enhanced message broadcast features, international roaming, and continued work with terminal, modem and application providers for PCMCIA radio modems, MOBITEX chipsets, and MOBITEX-enabled applications and software platforms.

References

1. Mobitex Interface Specification, Rev. 2A, RAM Mobile Data, February 1994.

2. Goodman, D.J. and Saleh, A. A. M., The near/far effect in local ALOHA radio communications, *IEEE Trans, Vehicular Tech.*, VT-36(1), 19-27, February 1987.

3   Arnbak, J. C. and Van Blitterswijk, W., Capacity of slotted ALOHA in Rayleigh fading channels, *IEEE J. Selected Areas Commun.*, SAC-5(2), 261-269, February 1987.

4. Namislo, C., Analysis of mobile radio slotted ALOHA network, *IEEE J. Selected Areas Commun.*, SAC-2, 583-588, July 1986.

17

Performance Analysis of Digital FM in Simulcast Environment

Rade Petrovic
Center for Telecommunications
University of Mississippi, University, MS 38677

Water Roehr
Telecommunication Networks Consulting
11317 South Shore Road, Reston, VA 22090

Dennis Cameron, Kunio Tano
MTEL Technologies
P.O. Box 2469, Jackson, MS 39225

Abstract: A simulcast digital FM system in the presence of Rayleigh fading and additive white Gaussian noise exhibits performance floors, where increased signal strength does not reduce the error rate. Results obtained both in the lab, using a radio channel simulator, and in the field show this performance. Effects of the speed of movement, transmitter frequency offset, maximum frequency deviation, differential propagation delay, and relative strength of simulcast signals are investigated. The explanation of the observed floors is found in the generalization of the "FM click" theory.

The results presented herein have been utilized for contributing to the definition of Narrowband Personal Communication System (NPCS).

1.   Introduction

Simulcast of digital FM signals is used in current paging systems (POCSAG, GOLAY), newly developed paging systems (ERMES, FLEX), and it is a candidate for Narrowband PCS. The goals of simulcast are to increase coverage area, improve building penetration, and reduce the effects of shading and multipath fading. Although there is a wealth of empirical data which illustrates the above gains, more research is needed to fully characterize the effects of different variables, such as speed of the receiver movement in a Rayleigh fading environment, transmitter frequency offset, maximum frequency deviation, relative strength of simulcast signals, etc.

Earlier papers [1-3] were mostly concerned with exploiting some controlled differences among the transmitted signals in both the time and the frequency domain. Recently a more comprehensive investigation has been published [4] that exhibits the error performance floors: situations where a residual error rate persists in the face of increasing signal strengths. There was no attempt to explain such results.

Here we report our investigations of the effects of different parameters on system performance in both weak and strong signal environments, and in the presence of Rayleigh fading and a simulcast signal. Further, we present a theory which explains the observed error floors, which is based on the generalization of the Rice's "click theory" [5-7].

2.   Experimental System Description

MTEL technologies is licensed by the FCC to operate an experimental narrowband (50 kHz) PCS system in the 930 MHz band. A modulation technique that uses long symbol intervals makes simple, low powered, portable units feasible [8-10]. The high bit rates are achieved by using M-ary (M > 2) modulation and multicarrier operation. In these investigations four subcarriers, 10 kHz apart, each modulated by four level FSK are used. The symbol rate is 3200 baud, giving the total bit rate of 25.6 kbps.

The peak frequency deviation f_d is kept smaller than the baud rate. Discriminator (frequency to voltage convertor) demodulation is used. Such systems, known as narrowband FSK, provide good performance with fairly simple receiver design [11].

The premodulation pulses are shaped with a 10 pole Bessel filter in order to reduce the signal spectrum width. In the lab test the transmitters were simulated by a battery of signal generators with combined outputs, while for the field tests we had designed two transmitters with approximately 32 W output power each. In the lab we simulated Rayleigh fading and simulcast using the HP11759B RF channel simulator.

The receiver consists of an RF down-convertor followed by A/D convertor, DSP processor, and a PC for data analysis and presentation. The DSP processing consists of IF filtering and channel separation, followed by a numerical discrimination, post detection filtering, clock extraction, and threshold comparison modules. As an output to the PC we obtain 8 bits per symbol. The PC detects the frame synchronization word and interprets the messages.

In the bit-error-rate (BER) tests the same message is repeatedly transmitted. The PC compares the received message to the stored template in order to determine the BER. We also assumed that the BCH (31,21) code, with parity bit, is used for error correction, allowing 2 errors to be corrected in a 32 bit word. Further we assumed the bit-interleaving of 32 words in order to disperse bursts of error. Based on this assumptions we calculated message-error-rate with the message being equal to one block of 32 interleaved words.

3.   Lab Experiments

Fig. 1 shows bit-error-rate vs. signal strength in Rayleigh fading environment for different Doppler frequencies. The Doppler frequency, f_D[Hz] is related to the speed of movement, v [km/h], at 930.9 MHz according to

$$v = 1.16 \, f_D \qquad (1)$$

The signal strength P[dBm] is related to energy-per-bit over nose density ratio E_b/N_o[dB] according to

$$E_b/N_o = 130 + P \qquad (2)$$

The system parameters used for the experiment are as follows:

Premodulation filter bandwidth 2.5 kHz.

- Peak frequency deviation 1.8 kHz, i.e. the frequency separation between symbols 1200 Hz.

- IF filter bandwidth 6 kHz (two pole Butterworth cascaded with four pole Bessel)

- Post detection filter bandwidth 2.4 kHz (two pole Butterworth)

Fig. 1 shows that the BER decreases with increased signal strength for weak signals, but it goes flat at the high signal level; this phenomenon we call the BER floor. Apparently the BER floor depends strongly on the speed of movement. Fig. 2 shows a similar diagram in the presence of a simulcast signal with the amplitude 3dB and 10dB lower than the main signal (The main signal amplitude is shown on the abscissa). The relative signal delay was kept zero, as well as the signal carrier offset. The tests were run for two speeds of movement, 3 km/h and 100 km/h. At the lower speed it is apparent that the simulcast increases the BER floor, more if the simulcast interference has a closer amplitude. At the higher speed the BER floor is virtually unchanged.

More detailed investigation showed that fast fading and simulcast interference cause independent error counts, and the total BER can be obtained by simple summation of their contributions. The simulcast error count increases as the differential delay increases. Also, the BER floor increases as the frequency offset between stations increases.

Fig. 3 show the message-error-rate (MER) vs. signal power in the same set of conditions as in Fig. 2. Note that the MER has the floor too, i.e. the BCH (32,21) code can not correct all the errors if the speed of movement is high enough. Further testing showed that the MER floor increases with reducing the amplitude imbalance, da, increasing the differential delay dt, and increasing the frequency offset between transmitters fo. Also, we found that an increase in frequency deviation reduces the error floors.

In order to investigate the source of errors we observed the output of the FM discriminator together with the FM signal by an oscilloscope. Fig. 4 shows channel simulator results for a single signal (monocast), -70dBm power, and fast Rayleigh fading (100 km/h). The lower trace shows the FM signal, with fading. The upper trace shows the 4 level signal at the output of discriminator. Note the sharp peaks coinciding with the deepest fading valleys. Those peaks we will call FM clicks.

Fig. 5 shows a similar diagram in the case of equisignal simulcast (da = 0 dB), fast fading (100 km/h), no differential delay (dt = 0), and small frequency offset (fo = 20 Hz). There is a large number of FM clicks of both polarities, again aligning with deep fading valleys in the FM signal.

Fig. 6 shows a zoomed portion of a single FM click occurring in a simulcast situation. The FM click does not exactly align with the minimum of the fading envelope, but this can be explained by the added delay in discriminator and post detection filtering. The nominal frequency of the FM signal is 29 kHz (obtained by two step down conversion from 930.9 MHz). The oscilloscope markers shown in the diagram show that the instantaneous frequency during the fade is smaller than the nominal one.

4.   Field Tests

Two experimental composite transmitters were built. Each has four subtransmitters, and each subtransmitter operates in the 4 FSK mode. The outputs of the subtransmitters ware

218

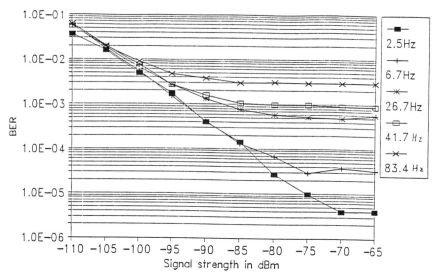

Fig. 1. Lab measurement of bit-error-rate us. signal strength for various
Doppler shifts.

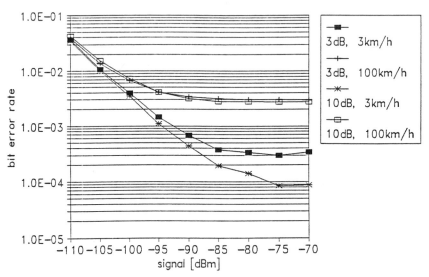

Fig. 2. Lab measurements of BER vs. signal strength in two signal simulcast
in Rayleigh fading environment for various amplitude imbalances, and
speeds of movement.

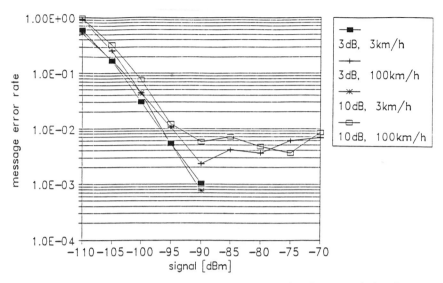

Fig. 3. Lab measurement of message-error-rate vs. signal strength in the same conditions as in Fig. 2.

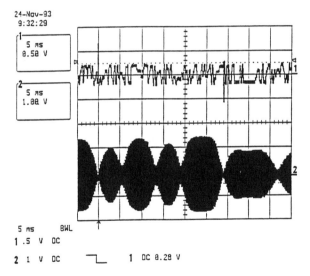

Fig. 4. Lower trace: Digital FM signal in Rayleigh fading environment.
Upper trace: Demodulated signal (four level FSK) with FM clicks marked with diamonds.

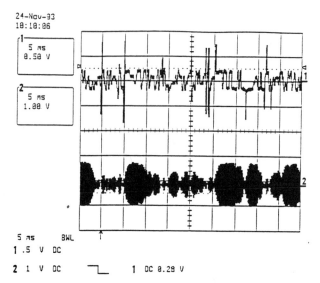

Fig. 5. Digital FM signal (lower trace) and discriminator output (upper trace) for simulcast (two signals nearly equal amplitude and frequency) in Rayleigh fading environment.

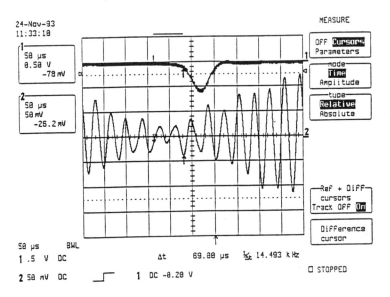

Fig. 6. Two signal simulcast with nominally equal amplitude and frequency (lower trace), and corresponding frequency discriminator output (upper trace). The time interval between markers correspond to two signal periods.

combined and broadcast by the same antenna. The composite transmitters were installed approximately 6 miles apart in rural, rolling terrain. The synchronization between transmitters was achieved by a GPS system. The receiver was installed in a van, with a whip antenna on the roof. For simulcast measurements we identified stretches of roads where the average amplitude imbalance between signals was less than 3dB.

Fig. 7 shows the measured BER vs. signal power [dBm] along one of the chosen road segments. The results obtained when a single transmitter was on (monocast) are marked by a square, and the results obtained in a simulcast environment (simul.) are marked by diamonds in the figure. The speed of the vehicle was about 30 mph. The dashed line shows lab result for monocast at this speed. The error floor is not so obvious as in the lab, in part due to a larger measurement error. Further, the Rayleigh fading model is not the best fit for a suburban area with roling hils, as in Oxford, MS. It can be shown that Rician model results in lower error floors.

The simulcast results show nearly the same BER as monocast. It should be noted that an improvement in BER in simulcast overlap area has been measured when longer runs are made. Then larger amplitude imbalances occur, and shadings block alternately one signal or the other.

Fig. 8 shows the effect of movement on BER floor. The results are obtained in a monocast environment using a different (slightly less sensitive) receiver. Evidently the higher speed results in the higher error count, which is in agreement with the lab results.

5. Analysis of Experimental Results

It is well know [5-7] that FM clicks occur in low signal-to-noise ratio (SNR) environment, when the signal plus noise vector in the phase diagram rotates around the origin. It is only natural to conclude that such events occur during deep fades, caused either by a multipath fading or by a simulcast interference fade. Increased speed of movement will increase the frequency of deep fades. Similarly frequency offsets and amplitude imbalances influence frequency and depth of simulcast fades. However, these effects are insufficient to explain the results we observed. First, by changing f_D from 2.5 Hz to 83.4 Hz (Fig. 1) the frequency of fades increased 33 times, while the BER increased 750 times. The fades are narrower at $f_D = 83.4$ Hz than at 2.5 Hz and equally deep, so they should not produce more errors. Secondly, the signal strength at the bottom of the fade should increase with the average signal strength, and BER should keep decreasing, i.e., the fade frequency hypothesis does not explain error floors. Thirdly, in simulcast without fading we observed FM click that all have the same polarity. By changing the sign of the amplitude imbalance, da, the polarity of FM clicks was inverted. The above observations lead us to the conclusion that the FM clicks are not caused by noise in the region of error floors. This can be verified by a simple mathematic analysis:

Our receiver has a numerical discriminator which calculates the output signal s(t) based on two input signals, in-phase x(t) and in-quadrature y(t), according to

$$s(t) = \frac{x(t)\dfrac{dy(t)}{dt} - y(t)\dfrac{dx(t)}{dt}}{x^2(t) + y^2(t)} \tag{3}$$

It is easy to show that s(t) is proportional to the input signal frequency and it is insensitive to

222

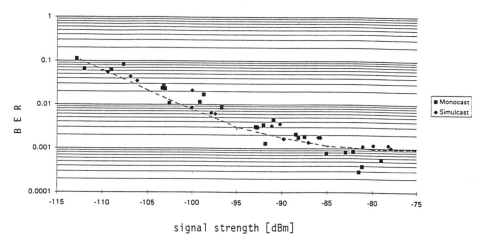

Fig. 7. Field measurement of bit-error-rate vs. signal strength for
single signal (squares) and two signal simulcast (diamonds).

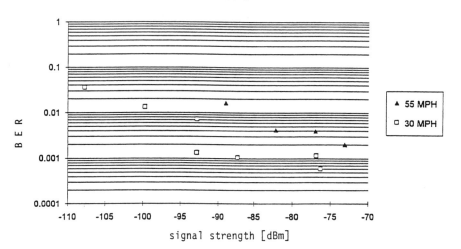

Fig. 8. Field measurement of bit-error-rate vs. signal strength for two
different speeds of movement.

amplitude modulation. Let us consider the input signals of the form

$$x(t) = a(t) (\cos \alpha(t) + c(t) \cos \beta(t))$$ (4)

$$y(t) = a(t) (\sin \alpha(t) + c(t) \sin \beta(t))$$

Where $a(t)$ represents a parasitic amplitude modulation, $c(t)$ is a time varying amplitude imbalance, $\alpha(t)$ and $\beta(t)$ are the phases of two simulcast signals. Alternatively they can be considered as phases of two rays in a multipath fading environments. By substituting (4) in (3), and after some algebra we obtain:

$$s(t) = \pi (f_a(t) + f_b(t)) + \frac{\pi (1-c^2(t)) \Delta f - \dfrac{dc(t)}{dt} \sin\Delta\phi}{1 + 2c(t) \cos\Delta\phi + c^2(t)}$$ (5)

where

$$f_a(t) = \frac{1}{2\pi} \frac{d\alpha(t)}{dt}$$

$$f_b(t) = \frac{1}{2\pi} \frac{d\beta(t)}{dt}$$

$$\Delta f = f_a(t) - f_b(t)$$

$$\Delta\phi = \alpha(t) - \beta(t)$$

The formula (5) produces signal with FM clicks peaking at the instances where $\Delta\phi = (2k-1)\pi$, $k = 0, \pm 1, \pm 2,...$ For example, in Fig. 9 we plotted both the discriminator input signal (lower trace) and output signal, $s(t)$, (upper trace) with the data similar to those observed experimentally in Fig. 6. A good match of analytic and experimental results is found.

For $\Delta\phi = (2k-1)\pi$ we can calculate, using (5), the amplitude of the clicks, A,:

$$A = \frac{1 + c(t)}{2} \frac{2\pi \Delta f(t)}{1 - c(t)}$$ (6)

Depending on $\Delta f(t)$ and $c(t)$, as well as on the current symbol, some clicks cause errors and some do not. But the error rate does not depend on the signal strength, $a(t)$, which explains error floors. In simulcast environment the increase in frequency offset increases both the frequency of fades (and clicks) and the click amplitude. Reduction of the amplitude imbalance increase the click amplitude. The polarity of the click is determined by the sign of A, and in a stable experimental environment (Δf and c are constant) all clicks have the same polarity, as observed experimentally.

In a multipath fading environment the frequency offset of participating waves is proportional to the speed of movement. Therefore with a higher speed, not only the frequency of fades increases, but also the amplitude of FM clicks. This explains the nonlinear dependence of the BER with the Doppler frequency observed in Fig. 1.

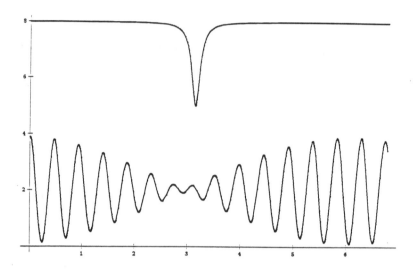

Fig. 9.   Computer simulation of two signal simulcast with nearly equal amplitude and frequency (lower trace), and corresponding output of a frequency discriminator (upper trace).

6.   Conclusion

Our experiments show that a digital FM with discriminator has a BER floor in a multipath fading or simulcast environment. This floor increases with speed of movement, frequency offset of transmitters, the differential propagation delay, and with decrease of amplitude imbalance of simulcast signals.

Earlier research [1] and empirical results suggest improvements in average BER with the increased frequency offsets. We have observed and analyzed error floors that increase (cause more errors) as the frequency offset increases. This paradox can be resolved by noting that error floor conditions (high signal levels and nearly matched signal levels) are relatively rare in field. In addition, in [1] wide band FM (with larger frequency deviation) was investigated, and a battery of filters wes used instead of a discriminator. Indeed our experiments show that the BER floor reduction can be achieved by increase in frequency deviation.

The explanation of the observed phenomena is found in a generalization of FM click theory. earlier papers used the theory to explain errors in low SNR environments. We have shown that FM clicks occur in the absence of noise if two or more signals interfere destructively. During a deep fade a small change in relative phase of participation waves results in a large and fast change of the composite signal phase, which generates a FM click. The amplitude of the click can be related to the frequency offset of the participating signals and their amplitude imbalance. Alternately it can be established that the clicks are higher if the fades are narrower and deeper.

References

[1]   T. Hattori, K. Hirade, and F. Adachi, "Theoretical Studies of a Simulcast Digital Radio Paging System Using a Carrier Frequency Offset Strategy", IEEE Trans. Vehic. Technol., vol. VT-29, pp 87-95, Feb. 1980.

[2]   T. Hattori and S. Ogose, "A New Modulation Scheme for Multitransmitter Simulcast Digital Mobile Radio Communication", IEEE Trans. Vehic. Technol., vol. VT-29, pp. 260-270, May 1980.

[3]   T. Hattori, K. Kaneko, and T. Nagatsu, "Multitransmitter Simulcast Digital Signal Transmission System Using Waveform Offset Strategy", Rev. Elec. Commun. Lab, vol. 30, no. 2, pp. 299-307, 1982.

[4]   F. Muratore, v. Palestini, and F. Pattini, "4-PAM/FM with Noncoherent Detection for a Pan-European Paging System" IEEE Trans. Commun., vol.41,no.11,pp. 1618-24, Nov. 1993

[5]   S.O. Rice, "Noise in FM Receivers", Chapter 25,pp. 375-424, in Proceedings, Symposium on Time Series Analysis, M. Rosenblatt (ed.) John Wiley & Sons, Inc., N.Y.,1963.

[6]   J.E. Mazo and J. Salz, "Theory of Error Rates for Digital FM", BSTJ, vol. 45, pp. 1511-35, Nov. 1966.

[7]   I. Korn, "Digital Communications", Chapter 8, pp.471-536, Van Nostrand Reinhold, 1985.

[8]   R. Petrovic, W. Roehr, and D. Cameron, "Multicarrier Permutation Modulation for Narrowband PCS", 3rd Virginia Tech. Symposium on Wireless Personal Communications, June 1993.

[9]   W. Roehr, R. Petrovic, and D. Cameron, "A Narrowband PCS Advanced Messaging System", ibid.

[10] R. Petrovic, W. Roehr, and D. Cameron, "Multicarrier Modulation for Narrowband PCS", IEEE Trans. on Vehic. Techn., accepted for publication.

[11] I. Korn, "M-ary Frequency Shift Keying with Limiter-Discriminator-Integrator Detector in Satellite Mobile Channel with Narrow Band Receiver Filter", IEEE Trans, on Com., Oct. 1990, pp. 1771-1778.

18

Improving Throughput and Availability of Cellular Digital Packet Data (CDPD)

Jay M. Jacobsmeyer
Pericle Communications Company
P.O. Box 50378
Colorado Springs, CO 80949

Abstract

Cellular Digital Packet Data (CDPD) is a packet radio system designed to provide reliable 19.2 kbps data service over the North American cellular phone system. The chief obstacle to high speed data transmission on CDPD is multipath fading. Time-varying amplitude and phase force CDPD to operate at a throughput far below that of equivalent non-fading channels. A new adaptive data rate technique can improve throughput on CDPD by more than a factor of 5. This technique adapts the data rate by varying the number of information bits, k, of a k/k+1 trellis code employing M-ary DPSK or CPFSK modulation. This adaptive trellis code ensures that we can raise the data rate without expanding the bandwidth. A novel soft-decision Viterbi decoder performs error correction, signal estimation, and rate changing in a single device. The built-in estimator automatically estimates the channel carrier-to-interference ratio with a standard deviation of 2.5 dB. An added benefit of the adaptive rate scheme is an increase in the service area of 17% for an omnidirectional cell and 9% for a directional cell.

1.0 Introduction

Cellular Digital Packet Data is designed to provide cost-effective data service over existing cellular radio channels. CDPD operates independently of the voice system, using either dedicated 30 kHz channels or hopping between idle channels. The CDPD channel, called the *channel stream*, is a shared medium employing digital sense multiple access (DSMA) with collision detection (CD). The system uses its own base station, called a mobile data base station (MDBS) and communicates with data mobiles, called mobile end systems (M-ESs). Calls are routed by intermediate systems (ISs), also called *routers*. CDPD is described in more detail in a specification [1], published in July 1993. Service will be available in several cities in 1994.

CDPD is a wireless channel and is therefore subject to a time-varying carrier-to-interference ratio (C/I). The two main causes of this time-varying C/I are Rayleigh fading and shadow loss.

2.0 Effects of Rayleigh Fading on Receiver Performance

Multipath fading on cellular channels causes time-varying amplitude and phase that seriously hamper high speed digital communications. To maintain acceptable bit-error-rates, CDPD must operate at an effective data rate that is far below channel capacity. In other words, a fading channel requires significantly greater power to achieve the same error rate performance as the equivalent non-fading channel.

To see why this is so, consider the Rayleigh fading mobile radio channel. Although the average signal energy of the fading channel is the same as the average signal energy of the

228

non-fading channel, the bit-error-rate (BER) for the Rayleigh fading channel is several orders of magnitude higher. For example, the probability of bit error for binary frequency shift-keyed (FSK) modulation on the non-fading additive white Gaussian noise (AWGN) channel is [6]

$$P_b = \frac{1}{2} e^{-\frac{E_b/N_0}{2}} \tag{1}$$

where E_b/N_0 is the ratio of energy per bit to noise power spectral density. But the probability of bit error for FSK on the Rayleigh fading channel is [6]

$$P_b = \frac{1}{2 + E_b/N_0} \tag{2}$$

These functions are plotted in Figure 1.

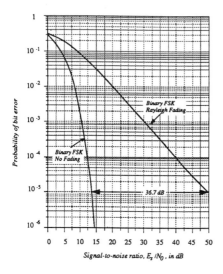

Figure 1 - Probability of bit error for AWGN and Rayleigh fading channels

Note that at a bit-error-rate of 10^{-5}, the Rayleigh fading channel requires almost 37 dB (5,000 times) more power than the non-fading channel.

Several techniques are used in an attempt to recover the performance lost to fading. The most common are worst case design, diversity, and error-correction coding. Worst case design requires large fade margins (sometimes greater than 40 dB) and is obviously inefficient. Diversity is a bit more effective (~20 dB improvement for dual diversity), but often impractical. Frequency diversity requires additional spectrum, a valuable resource that is usually not available, and space diversity is impractical on handheld radios.

A third approach is error correction coding. CDPD employs a good burst-error correcting code, the Reed-Solomon (RS) code. The particular code used is a RS (63,47) code over GF(64). This code can correct up to 8 symbol errors (a symbol is 6 bits), but the CDPD specification [1] recommends that the code correct no more than 7 errors so the remaining parity symbols can be used to provide better error detection performance. Since the Reed-Solomon code provides both error correction and error detection, a cyclic redundancy check (CRC) is not required.

CDPD employs Gaussian minimum shift keying (GMSK), a binary modulation technique. Most mobile receivers will use non-coherent detection, and the error performance will be approximately the same as non-coherent binary FSK.

Although the Reed-Solomon code is a powerful burst-error correction code, it is not failsafe, and many Reed-Solomon blocks will be uncorrectable because the fading channel will introduce more than 7 symbol errors. To handle this problem, CDPD will implement an automatic repeat-request (ARQ) scheme and will re-transmit uncorrectable blocks. The CDPD protocol makes it impractical to retransmit individual blocks, so an entire frame must be retransmitted if one or more Reed-Solomon blocks contain uncorrectable errors. A CDPD frame holds a minimum of 4 Reed-Solomon blocks and the protocol behaves like a go-back-N ARQ system with $N = 4$. From [10], we know that the throughput for a go-back-N CDPD system is approximated by

$$R_{avg} = (19.2 \text{ k}) \frac{P_{cd}}{P_{cd} + (1 - P_{cd})N} \left(\frac{k}{n}\right)$$

where P_{cd} is the probability that a RS block is correctly decoded, and (k/n) is the code rate of the RS code (3/4 in our case).

Computing P_{cd} for a Rayleigh fading channel is no simple task because of the strong time correlation of symbol errors. We resorted to simulation with a software simulator meeting the requirements of TIA IS-55A [11].

The results are shown in Table 1. Note that at a typical cellular C/I of 17 dB, CDPD produces less than half of its rated throughput.

Table 1 - Go-Back N ARQ Performance of CDPD

E_b/N_0	P_{cd}	R_{avg}
7.0 dB	0.06	240.0 bps
9.5	0.21	880
12.0	0.41	2,160
14.5	0.61	4,020
17.0	0.83	7,930
19.5	0.92	10,540
22.0	0.96	12,340

Table 1 serves as a baseline for CDPD performance and it will be compared in Section 4.0 to adaptive data rate performance.

3.0 Effects of Shadow Loss and Co-Channel Interference on Cell Coverage

Cellular radio is a frequency reuse system and its performance is usually limited by co-channel interference rather than Gaussian noise. The main source of this co-channel interference is the first tier of co-channel cells, shown in Figure 2 for a 7-cell reuse pattern.

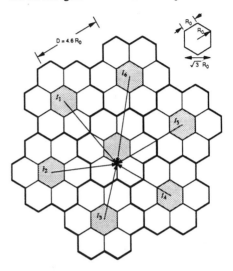

Figure 2 - Co-Channel Interference in a 7-Cell Reuse Pattern

As a mobile user travels through the cell, he or she will experience a time-varying carrier-to-interference ratio caused by terrain and made-made obstructions. This loss is known as *shadow loss*. At cellular frequencies, shadow loss varies about 100 times more slowly than Rayleigh fading. The main effect of shadow loss on a conventional cellular system is a loss of service in certain regions of the cell due to low C/I.

To quantify this effect, we need a relevant measure of the C/I in co-channel cells. The most appropriate measure is the probability that a randomly selected point in the cell of interest (the center cell in Figure 2) will have a C/I of u_{min} dB or greater. In other words, we want the probability that a mobile user has a $C/I \geq u_{min}$ dB, assuming the mobile user's position is uniformly distributed over the cell's area. Typical values of u_{min} for analog voice are 17 or 18 dB.

The following derivation of this probability is similar to Lee's derivation in [2], but we think our derivation is easier to understand.

The probability $P(U \geq u_{min})$ is written as

$$P(U > u_{min}) = 1 - P(U \leq u_{min})$$

$$= 1 - \int_0^{R_0} P(U \leq u_{min} \mid r) f_R(r) \, dr \tag{3}$$

where

$f_R(r)$ = probability density function for the distance from the base station to the user location, R, and
R_0 = the radius of the cell

From [9, pp. 191], we know that the probability density function for R is given by

$$f_R(r) = \frac{2r}{R_0^2} \tag{4}$$

where we have approximated the cell as a circle of radius R_0.

The conditional distribution of U, $P(U \leq u|r)$, is found as follows: Because of terrain shadowing and man-made obstacles, the carrier, C, and the interference, I, at any point in the cell are random variables. One of the most widely accepted probability distribution for C and I is the log-normal distribution [2], [4]. Therefore, if C and I are in dB, they are normally distributed with mean c_0 and i_0 and variance σ_C^2 and σ_I^2, respectively. But we are interested in the random variable, $U = C - I$.

From [9], we know the the sum of two independent normal random variables is also a normal random variable with mean equal to the sum of the means and variance equal to the

sum of the variances. Therefore, the carrier-to-interference ratio, U, is normally distributed with mean $c_0 - i_0$ and variance

$$\sigma^2 = \sigma_C^2 + \sigma_I^2 \tag{5}$$

For convenience, we can write the probability $P(U \leq u|r)$ in terms of the unit normal distribution, $\Phi(x)$,

$$P(U \leq u_{min} \mid r) = \Phi\left(\frac{u_{min} - u(r)}{\sigma}\right) \tag{6}$$

where

$$\Phi(x) = \frac{1}{\sqrt{2\pi}} \int_{-\infty}^{x} e^{-\frac{y^2}{2}} \, dy \tag{7}$$

and $u(r)$ is the mean C/I at a distance r from the base station. Assuming that the interference is roughly constant throughout the cell [4], the mean C/I at a distance r from the base station can be written as

$$u(r) = u(R_0) - \gamma \log\left(\frac{r}{R_0}\right) \tag{8}$$

where γ is the propagation constant (typically $\gamma = 40$). Substituting (8) into (6) and (4) and (6) into (3), we get

$$P(U > u_{min}) = 1 - \frac{2}{R_0^2} \int_0^{R_0} \Phi\left(\frac{u_{min} - u(R_0) + \gamma \log\left(\frac{r}{R_0}\right)}{\sigma}\right) r \, dr$$

Making a change of variables, $s = r/R_0$, we arrive at the following expression:

$$P(U > u_{min}) = 1 - 2 \int_0^1 \Phi\left(\frac{-\Delta u + \gamma \log(s)}{\sigma}\right) s \, ds \tag{9}$$

where $\Delta u = u(R_0) - u_{min}$. This expression can be integrated using numerical integration.[1] Equation (9) is plotted in

[1]The function has a discontinuity at $s = 0$ which will cause most numerical integration routines to fail.

Figure 3 (at the end of this paper) for $\sigma_c = 6$ dB (small cell) and $\sigma_c = 8$ dB (large cell).

Next, we need the value of $\Delta u = u(R_0) - u_{min}$ for an omnidirectional cell site. Assuming a propagation constant of $\gamma = 40$ (i.e., 40 dB per decade path loss), the carrier power at the edge of the cell is proportional to the cell radius to the minus fourth power,

$$C \propto R_0^{-4}$$

And the carrier-to-interference ratio is

$$\frac{C}{I} = \frac{R_0^{-4}}{\sum_{i=1}^{6} d_i^{-4}}$$

(10)

where d_i is the path distance from the center of cell i to the edge of our cell. Referring to Figure 2, we can calculate the path distances to the edge of our cell. These distances are shown in Table 2.

Using these distances and (10), we find that the C/I at the edge of the cell is 17.8 dB. Therefore, Δu is 0.8 dB for a $u_{min} = 17$ dB. Note from Figure 3 that this results in a coverage of 74.5%, which is far less than the usual design goal of 90%.

Table 2 - Co-Channel Cell Distances

Cell, i	Distance, d_i
1	$5.3 R_0$
2	$4.4 R_0$
3	$3.6 R_0$
4	$4.0 R_0$
5	$5.0 R_0$
6	$5.6 R_0$

We recommend the midpoint rule. This expression can be integrated analytically, but the derivation is omitted in the interest of brevity. The analytical result is in terms of the error function, so numerical integration or approximation is required regardless.

Partitioning a cell into 3 sectors with 120° beamwidth antennas will increase the C/I at the cell boundary. Assuming perfect antennas, only cells 1 and 2 of Figure 4 contribute to co-channel interference.

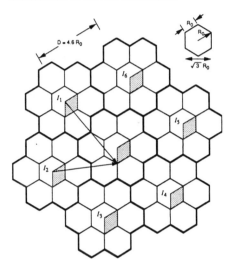

Figure 4 - Co-Channel Interference with Directional Cells

Using (10), we can show that sectorizing increases the C/I at the cell boundary by over 6.2 dB. From Figure 3, we see that this increase in C/I raises the area covered with greater than $u_{min} = 17$ dB to 87.5%.

We will show in the next section that an adaptive data rate modem lowers u_{min} and therefore increases the coverage area over a conventional CDPD modem.

4.0 Adaptive Data Rate Performance

Our approach is to adapt the system data rate to the time-varying signal-to-noise ratio at the mobile radio receiver. To understand why this approach makes sense, we must re-examine the fading characteristics of the mobile radio channel.

232

The vast majority of bit errors on mobile radio occur during deep fades. As the signal-to-noise ratio drops below the system threshold, the probability of bit error rises rapidly and errors are clustered in bursts coinciding with the fade. Even in the absence of intersymbol interference, burst errors still occur because the data rate is too high for the instantaneous signal-to-noise ratio. In other words, the energy-per-bit is too low.

The most obvious way to increase energy-per-bit is to boost the transmitter power. But power is expensive, and most systems already operate at the highest practical transmitter power. The next logical choice is bit rate. Since $E_b = P_t/R$ (where P_t is the transmitter power in watts and R is the bit rate in bits per second), we can double the energy-per-bit simply by halving the bit rate. An adaptive data rate system exploits this phenomenon by operating at high bit rates when signal-to-noise ratios are high, and low bit rates when signal-to-noise ratios are low. But whenever we lower the bit rate, we degrade throughput. So the obvious question is: Do we get a net gain or a net loss? The answer is a net gain, a fact that was proved mathematically by Bello and Cowan in 1962 [7].

For example, consider the simplest adaptive data rate scheme: one that operates at rate R bits/s when the received signal is above a predetermined threshold, and 0 bits/s when the received signal is below this threshold.

To quantify the performance of this scheme, we recall that the power on a Rayleigh fading channel is exponentially distributed [6] and assuming that the minimum required signal-to-noise ratio is also the mean signal-to-noise ratio, the probability that we drop below this minimum is simply

$$P = 1 - e^{-1} = 63\%$$

Assuming ergodicity[2], we can conclude that we lose only 63% of our throughput, which is equivalent to a 4.3 dB loss in power over a Gaussian noise channel. Or put another way, we gain more than 32 dB over conventional FSK in Rayleigh fading. We see this result plotted in Figure 5 below.

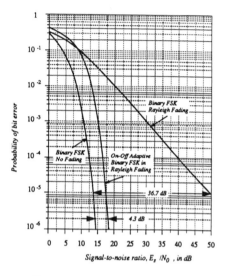

Figure 5 - Performance of a simple on/off adaptive data rate system

To achieve this result, we need a system that operates at a constant data rate of R bits per second (bps) for 37% of the time and drops instantaneously to 0 bps at the onset of a deep fade. Of course, accurate and timely channel estimation is required to approach this performance in a practical system.

Next consider a more sophisticated adaptive data rate system with n available non-zero rates, $r_1, r_2, \dots, r_n$. Each rate has a corresponding signal-to-noise ratio per bit, $(E_b/N_0)_i, i = 1, 2, 3, \dots, n$ These signal-to-noise ratios are the thresholds that

2An assumption backed by experimental evidence, see Lee [4].

determine when the system can raise or lower the data rate and still maintain reliable communications. We normalize the adaptive system by letting

$$a_i^2 = \frac{(E_b/N_0)_i}{(E_b/N_0)_{mean}}$$

Then we can write the probability mass function for R for an n-rate system as

$$p_R(i) = P(R = r_i) = (e^{-a_i^2} - e^{-a_{i+1}^2}),$$

$$i = 1, 2, 3, ..., n-1$$

$$p_R(n) = P(R = r_n) = e^{-a_n^2}$$

And the average bit rate for an n-rate system on the Rayleigh fading channel is simply

$$R_{avg} = \sum_{i=1}^{n-1} r_i (e^{-a_i^2} - e^{-a_{i+1}^2}) + r_n e^{-a_n^2} \tag{11}$$

Now, what type of adaptive rate scheme should we use, and what are the possible rates and thresholds?

We could design a system that increments the channel symbol rate by a factor of 2 whenever the signal-to-noise ratio increases or decreases by 3 dB, but such a system would require bandwidth expansion and the CDPD channel is bandlimited.

A better approach is to use a constant symbol rate and adapt the data rate by varying the number of bits per channel symbol. We should also employ trellis-coded modulation (TCM) to improve the power efficiency of the system.

The E_b/N_0 thresholds for a 3-rate adaptive TCM scheme with M-ary DPSK are shown in Table 3 below.

The allowable bit rates are 1 bit/symbol (coded DQPSK), 2 bits/symbol (coded 8-DPSK),

and 3 bits/symbol (coded 16-DPSK).

Table 3 - Signal-to-noise Thresholds for Adaptive TCM (BER = 10^{-5})

Modulation	Code Rate	Required E_b/N_0
DQPSK	1/2	8.3 dB
8-DPSK	2/3	10.5 dB
16-DPSK	3/4	14.4 dB

The improvement in throughput of the trellis-coded adaptive rate scheme over fading binary FSK is dependent on the mean signal-to-noise ratio on the channel. A typical cellular signal-to-noise ratio is 17 dB. Substituting the rates and signal-to-noise ratio thresholds from Table 3 into Equation (11), we get

$$R_{avg} = 2.3 R = 44.2 \text{ kbps} \tag{12}$$

for a mean E_b/N_0 of 17 dB where R = the CDPD signaling rate of 19.2 kHz.

From Table 1, we see that the improvement over binary FSK with Reed-Solomon error correction is a factor of 5.4.

The throughput of this adaptive data rate technique is compared to conventional CDPD as a function of signal-to-noise ratio in Figure 6.

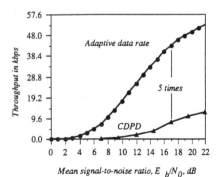

Figure 6 - Adaptive data rate and CDPD throughput (f_c = 881 MHz, velocity = 30 mph)

234

Note that the adaptive rate improvement is a decreasing function of signal-to-noise ratio. For example, the gain at 7 dB is a factor of 30, but the gain at 7 dB is factor of 5.4.

An added benefit of adaptive data rate is a gain in geographical coverage. Let's assume that a conventional CDPD modem requires a C/I of 17 dB to provide reliable service. In Table 3, we showed that an adaptive data rate system requires only 8.3 dB E_b/N_0 or C/I to operate at its lowest data rate.[3]   For an omnidirectional cell with 17.8 dB C/I, $\Delta u = 9.5$ dB. From Figure 3, we see that such a system can raise the area covered by at least u_{min} C/I from 74.5% to 91.2%. For a directional cell (3 sectors), the adaptive rate system raises the service region from 87.5% to 96.8% of the cell's area.

5.0   Implementation Issues

5.1 Introduction. Adaptive data rate schemes make sense from a theoretical view, but can high performance adaptive data rate modems be realized in practice? We think so.

Pericle Communications Company is developing an adaptive data rate modem for cellular radio.[4] The modem combines trellis coded modulation (TCM) and adaptive data rate operation in a single device. The trellis encoder operates at a constant symbol rate and varies the bit rate by adapting the number of bits per channel symbol to the channel conditions. The code rate of the encoder is $k/k+1$, $k = 1, 2, 3$. The encoder is a constraint length 3 convolutional code. To date, we have investigated M-ary DPSK modulation, but the modem architecture will accommodate other bandwidth efficient modulation techniques such as M-ary continuous phase frequency shift keying (CPFSK).

The rate changing algorithm and signal estimator are integral parts of a soft-decision Viterbi decoder, simplifying the implementation and reducing size.

We are addressing several implementation issues, including the following:

- Mobile access and overhead losses
- Signal estimation
- Memory & memory control
- Feedback channel effects
- Irreducible error floor

5.2 Mobile Access. CDPD is a shared medium. When a mobile is monitoring the system before access, CDPD is a one-to-N system. When a mobile gains access to the channel stream, the exchange of data between the mobile and the base is a one-to-one system. Busy/idle status, synchronization, and other control information are interleaved with the data, but this information occurs in separate channel symbols. Our modem operates at a constant symbol rate and varies the number of bits per channel symbol to vary the information rate. With proper framing information, the modem can change the data rate on a symbol-by-symbol basis.

Thus, we can send control information at the lowest non-zero rate of 1 bit per symbol, but the user information can be sent at a variable rate of 0, 1, 2, or 3 bits per symbol. For example, consider a generic packet radio system with user data preceded and followed by control information as shown in Figure 7.[5] In the discussion that follows, we will refer to the forward link (base to mobile). The modem works equally well on the reverse link (mobile to base).

---

[3]Assuming the interference is Gaussian distributed.
[4]Patent pending.

[5]This is not how CDPD works. We are using a simplified scheme to aid the explanation.

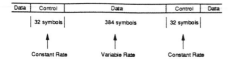

Figure 7 - Frame Structure Showing
Variable Rate Transmission

In this example, we have 32 symbols of control information preceding each 384 symbol block of user data. Within the block of user data, we can adapt the data rate almost continuously to the instantaneous channel conditions. Of course, we require a return path (i.e., a full duplex link) to pass rate information to the base station.

If the data transfer is one-way, the entire return path can be used for passing rate information. In this case, we could theoretically change the rate every symbol, but some time delay is necessary to transmit and receive the rate information sent on the return path. A minimum delay of 3 symbols is more realistic.

If the data transmission is two-way, rate information must be included as overhead in the user's data stream. One way to do this is shown in Figure 8.

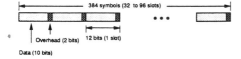

Figure 8 - Slot Structure Within A Data Block

Note from Figure 8 that the data block carries between 32 and 96 slots where each slot comprises 10 data bits and 2 overhead bits. Only 2 overhead bits are required since we have only four possible rates. The overhead is 17%, which is less than the overhead required by the Reed-Solomon code. Since we will use trellis-coded modulation, the Reed-Solomon code would not be required for

error correction. For error detection, the Viterbi decoder may be used (although this gets complicated) or we may use a 16-bit CRC. Details such as these are beyond the scope of this paper. The main point is that we can adapt the rate dynamically to accommodate a one-to-N packet radio system.

5.3 Signal Estimation. An adaptive data rate system also requires an accurate and timely estimate of the received signal-to-noise ratio. In the system under development, the estimator is built into the Viterbi decoder and it estimates signal-to-noise, not just signal. An example of the estimator's operation is shown in Figure 9 at the end of this paper. At typical cellular doppler frequencies, the estimator tracks the signal well with a standard deviation of 2.5 dB. Since the Viterbi decoder can correct many error patterns, an infrequent bad estimate does not necessarily result in a bit error. In Figure 9, we see several cases where the estimator overestimated the signal, but no error occurred.

When the signal is below the threshold for our lowest data rate (1 bit per symbol), information transfer ceases and training symbols are sent. The estimator continues to measure the channel and directs the transmitter to resume information transfer as soon as the signal rises above threshold.

5.4 Memory. For a conventional constant data rate system, the delay to receive an N-bit packet is simply $T = N/R$ where R is the data rate in bits per second. When a constant data rate system employs a block error correction code or an interleaver, the delay is increased by the encoding time or interleaver depth, δ, so that the total delay is $T = \delta + N/R$. In either case, the delay is a constant, independent of the channel conditions.

Adaptive data rate systems are different. Packet delay is a random variable that is a function of the amplitude and doppler characteristics of the channel. Consequently, additional memory is needed to provide a buffer between a constant data rate source and the variable data rate modem. On CDPD, overflow of the buffer is not a problem provided good control is used. CDPD already assembles packets in the mobile terminal's memory and can feed one packet at a time to the adaptive rate modem.

5.5 Feedback Channel. Another obvious obstacle to adaptive data rate communications is the need for a reliable feedback channel. On mobile radio, the feedback channel will also fade and will therefore introduce a time-varying signal, noise and delay.

In the modem, a reliable data rate is estimated by the channel estimator and a word, K_i, representing the desired data rate, is transmitted on the feedback channel to the forward channel transmitter. Propagation delay adds a 2 symbol delay to the processing time. An additional delay is built in to the estimator since it must have a non-zero sample size, n, to make an estimate.

The data rate word is subject to fading and noise and errors will occur in accordance the channel statistics. When an error occurs on the feedback channel, the transmitter and receiver are operating at different data rates and a loss of rate synchronization occurs. When rate synchronization is lost, the bit-error rate is almost 1/2, and a packet transmitted during this time will almost certainly require re-transmission.

All of these feedback channel effects have been simulated. Simulations show that the ultimate effect of the feedback channel is a loss in throughput of approximately 15% over theoretical.

5.6 Irreducible Error Floor. On fading channels, differential PSK modulation suffers from an irreducible error rate caused by the random phase rotation that occurs during a symbol period. The extent of the problem depends on the symbol rate and the doppler frequency of the channel. We would prefer low doppler and high symbol rates to minimize this problem. One advantage of an adaptive data rate system is that it does not transmit information during the deep fades when the random phase rotations are greatest. Consequently, the irreducible error floor for an adaptive data rate system will be lower than for a conventional system.

6.0 Conclusions

CDPD is a wireless communications system and is therefore subject to the time-varying signal levels that are inherent to wireless channels. Actual throughput on CDPD will be less than half the rated throughput of 19.2 kbps because of Rayleigh fading. We can improve the throughput of CDPD by a factor of 5 by adapting the data rate to the instantaneous signal-to-noise ratio at the radio receiver. An adaptive data rate system will also increase the service area of a cellular base station by over 17%.

Pericle Communications Company is developing an adaptive data rate modem for cellular applications. The modem combines rate changing, signal estimation, and error correction in a single device. An efficient decoding algorithm ensures the modem can be built cost-effectively.

7.0 Acknowledgement

Dr. Mark A. Wickert developed the Rayleigh channel simulator used to generate the CDPD performance measures used in this paper. Dr. Wickert is an associate Professor with the Electrical and Computer Engineering Department at the University of Colorado at Colorado Springs.

8.0 References

[1] Cellular Digital Packet Data System Specification, Release 1.0, July 19, 1993.

[2] W.C.Y. Lee, "Data Transmission via Cellular Systems," 43rd IEEE Vehicular Technology Conference, May 18-20, 1993, Secaucus, New Jersey.

[3] V. H. MacDonald, "The cellular concept," *The Bell System Technical Journal*, January 1979.

[4] W.C.Y. Lee, Mobile Cellular Telecommunications Systems, New York: McGraw-Hill, 1989.

[5] N. J. Boucher, The Cellular Radio Handbook, 2d Ed., Mendicino, CA: Quantum, 1992.

[6] R.E. Blahut, Digital Transmission of Information, Reading Massachusetts: Addison Wesley, 1990.

[7] P.A. Bello and W. M. Cowan, "Theoretical study of on/off transmission over Gaussian multiplicative circuits," *Proceedings of the IRE National Communications Symposium*, Utica, NY, Oct 1962.

[8] J. K. Cavers, "Variable-rate transmission for Rayleigh fading channels," *IEEE Transactions on Communications*, vol. COM-20, no. 1, pp. 15-22, Feb. 1972.

[9] S. M. Ross, A First Course in Probability, New York: MacMillan, 1984.

[10] S. Lin and D.J. Costello, Error Control Coding Fundamentals and Applications, Englewood Cliffs, NJ: Prentice Hall, 1983.

[11] TIA IS-55A, "Recommended Minimum Performance Standards of 800 MHz Dual-Mode Mobile Stations," September 1993.

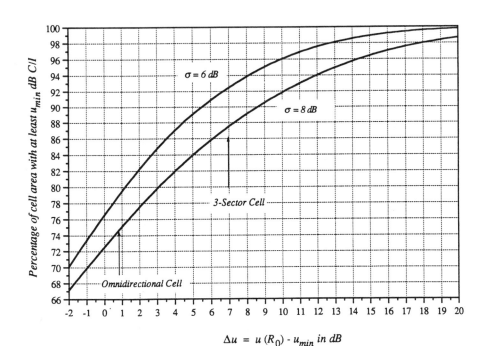

$$\Delta u = u(R_0) - u_{min} \text{ in } dB$$

Figure 3 - Probability that $C/I > u_{min}$

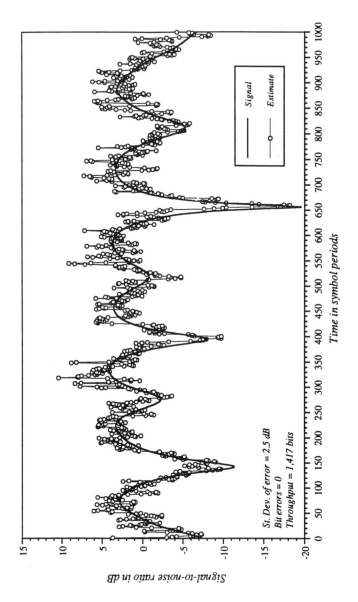

Figure 9 - Estimator Performance

The Impact of Defense Conversion on the Wireless Industry:

Corporate and Government Perspectives

Jeffrey H. Reed
Mobile and Portable Radio Research Group
Virginia Tech
Blacksburg, VA 24061-0350
Internet: reedjh@vtvm1.cc.vt.edu

The 1994 Virginia Tech Symposium on Wireless Personal Communications included a very special panel session on the impact of defense conversion on the wireless industry. The panel included Randy Katz from ARPA, Henry Bachman from Hazeltine, Joe Kennedy from ERA (an E-Systems company), Jerry Woodard from Signal Science, and Bruce Cole from ESL, all of whom are actively involved in transitioning defense technology to wireless communications products. They discussed the opportunities and pitfalls awaiting defense contractors entering the commercial wireless business.

The strategies for successful transitioning defense technology to the commercial wireless business varied greatly among the panel participants. Jerry Woodard stressed the need to completely .desolate the old corporate structure geared to defense business and build a new structure and corporate personality. Henry Bachman and Joe Kennedy stressed focusing on commercial market areas that capitalize on their defense skills and technology expertise. Bruce Cole said that ESL's strategy was to spin-off companies composed of ESL employees that focus on specific niche market areas using technology developed in ESL's military programs. Randy Katz provided his insight into viable commercial technologies from his prospective as a reviewer of ARPA's Technology Reinvestment Program (TRP) proposals. This is a program designed to transition defense technology to commercial markets. All panel participants agreed that relief from government accounting procedures and regulations would be the most valuable help that the government could provide to contend with defense cutbacks.

19

ARPA'S Prospective of Defense Conversion Opportunities for the Wireless Industry

Randy Katz
ARPA
3701 N. Fairfax Drive
Arlington, VA 22203-1714

The plan for the panel session was for me to give an overview from the ARPA prospective of where we see the opportunities in wireless communications, and some of the things that have been happening in defense conversion and in the ARPA managed technology reinvestment project, which is the main technical thrust of defense conversion. Then we will turn it over to the panel members to make their own pitches, somewhere around a 5 minute time frame, about their views on opportunities and challenges in defense conversion for wireless communications. By-the-way, if you do have some good ideas on defense conversion, we are still collecting topics, so I encourage you to try and contact me if there is some wild and crazy idea you have in the course of this panel session.

Lets put things in something of a prospective here. The defense mission is kind of unique, and part of the difficulty that we have with defense conversion or the application of commercial wireless technologies to defense, is that there are certain aspects of what the DoD needs to do that are not the same as pedestrians walking around using PCS systems. In defense, we have to worry about individuals and other platforms, some of which will be relatively slow, to 10's of kilometers per hour, some which move very fast, maybe 100's of kilometers per hour. We need to be able to communicate to all of them, and they need to be integrated in a reasonable fashion; a single command and control type of environment is necessary. We would like to be able to communicate with those people, platforms, entities anytime, anywhere, which is where wireless comes in as a critical need. The environment tends to be very hostile. You have adversaries who are trying to assure that you cannot communicate amongst those platforms. Although you can draw some analogies between interference and active jamming, they are very different problems, and that is another one of the issues about whether you can take commercial technologies and apply them.

It turns out that defense, within the Federal Government, has been one of the key patient and long term investors in communications technologies. Defense gets the credit for things like spread spectrum technology, frequency hopping, and antennae arrays. There is a whole slew of innovations over the past 50 years, and probably even earlier than that, that has been written by the military. The questions is whether a lot of those spin-offs can be brought to commercial fruition, to help those defense companies which have participated in creating that technology. In addition, are there some commercial technologies which have spun-off along the way in communications, that can be mapped into making more portable, more capable communication systems for the military? The challenge we face everywhere within the Department of Defense today, the underlying theme, is an integrated defense and commercial technology base. You don't want to have specialized equipment and you want to minimize the kind of specialized equipment you need for the defense mission. You would like to be able to use 90% commercial technology, 10% defense specific technology for example. We simply cannot afford to continue to have defense only systems for communications or anything else. As part of that, given the shrinking and the size of the defense budget, the first thing that always gets cut when you shrink the defense budget is the procurement and investments in research and development. You can't get rid of the people or shrink the size of the forces all that fast. So your portable communication systems will be very important, especially taking commercial technology and see how it can be applied mostly in the defense realm, realizing that there probably will be some capabilities that will need to be retrofitted and added. You need communication products and services at competitive prices for the commercial sector. Can we take some of that defense technology and yield world beating technologies that will be of interest to commercial users and consumers.

One of my key stories about the way the DoD works is that they will take these incredible capable multi-mode frequency agile radio systems which fit on the back of a truck, invest millions of dollars in an R&D program and manage to shrink that system into an 80 pound backpack, and this is what they call a portable communication system. What we really need to do is to look at how we can take things like Motorola Microtec communicators and add, for relatively modest weight - power and capability - the kinds of needs the DoD has. I believe a very different mind set is necessary to do that.

Now if you will look at the requirements, the DoD is still very much interested in anti-jam, low probability of intercept and detect. Interestingly enough, there is probably more in common with the needs of DoD and law enforcement agencies than there ever could be between the DoD and commercial or consumer users of communications technologies. But the things we need are the usual ones, make it hard to find me, make it hard to tell that I am using the communicator, make it hard to jam me, make it possible to use the available spectrum in a very flexible manner. You may want to harness that spectrum for the small number of high bandwidth channels or a large number of low bandwidth channels or enormous number of very, very low bandwidth channels along the lines of the in- situ sensors that Ken Gabriel talked about at lunch today. We need an enormous amount of flexibility in the way the topology of the network is created, the old radio problem. Also, the routing flexibility, quality of service, kinds of flexibility that has to be able to work in a variety of different environments for a variety of different needs and has to be able to change how it is behaving on the fly. That is critical for the military needs. We need modeling and tools for designing and managing these networks. The kinds of things that, for example, are done at Virginia Tech, would be extremely useful in the defense realm. Another key mission for the Department of Defense, as we move into the new world order, is the need to establish communications infrastructure very rapidly, so called the "instant infrastructure". This is the ability to use overlay technologies so that when you roll into Ethiopia, Somalia, or Bosnia, you can establish the communications capability very rapidly in a country which may have its own communications systems which you can't take advantage of, or may have no communications at all. So technologies that can do that, which are not that different from reestablishing communications after a local natural disaster, are the kinds of things the Department of Defense needs. One of the lessons learned in Desert Storm was that the communications infrastructure could not keep up with the troops. So it is also possible to have a real communications infrastructure that can follow along with the troops as they advance in the field.

Now that we have already had one round in the Technology Reinvestment Project (TRP), I thought you would be interested in knowing who the winners were in the wireless communications area. I was the primary coordinator for those proposals. Over 100 of the roughly 2800 proposals that came into TRP, came into the wireless topic area. This wasn't a very well written topic. I must tell you that one of the things that you could propose was terahertz optical fibers. The only reason that I can think of this as being part of wireless communications technologies is that the signals are not going over copper wires. But, nonetheless, there were several different technologies that were mentioned. The winners are shown here and it gives you some appreciation for the sizes of the winning proposals. One very interesting project is using what they called "spatial division multiple accesses". It actually uses kind of an old idea of using advanced antenna and signal processing techniques to be able to track the location of terminals, and do some antenna beam forming and steering, so that you could basically reuse frequencies within a sector if you could track the location of the terminals as they move around within your sector. They are in the process of trying to commercialize this technology and build it into base stations. The technology, the details of making it work on the fly a truly mobile environment, were actually developed under the Star Wars program for tracking "you know what's". Another project was actually using spread spectrum communications techniques to control BART (bay area rapid transit) trains inside tunnels. A couple of projects focus on basic technology for the use in higher frequency communications systems. A truly cool project, I thought was the TRW project, the passive millimeter wave camera. A camera that uses millimeter wave technology to be able to see through smoke and fire. It has potentially an enormous value to firefighters and people dealing with natural disasters.

These are some of the topic areas that we are interested in, and I will go through these very quickly. We are interested in super smart radio technology that can be developed in a cost effective or cost reduced fashion. Multi-band, multi-mode radios that can be very agile with the waveforms they can process and the kinds of schemes they can use, so that they can be used anywhere and adapt to the environment that you are carrying them into. We are also interested in building on top of the commercial base, the network and developments of things like the mobile Internet Protocols, to start seeing the development of commercial products built around that emerging standard from the Internet community. I think it is very important to look beyond voice as the critical application in communication systems. The DoD, and I believe many other users, are interested in wireless data communications as well as voice communications. So, a critical aspect is how you integrate the wireless data communications world with the wireline networked infrastructure that you might have, including the access to servers and services within that wide infrastructure. Further network protocols have to be developed for mobility, right now it is a very hokey system where the Mobile IP protocols are really based on a very slow re-deployment of assets. It's very slow, I take my computer from Washington down to Virginia Tech and I start using it down here rather than rapidly re-deploying it through a cellular system that I might be working within. And we are also interested in what I guess Ken Gabriel would call "body local area networks -- body LANS" so that you can use that to interconnect various equipment that a person might carry.

The last thing I want to cover, since this is a university sponsored symposium, is that I think that universities have a very important role to play in this conversion process, because universities bring a very interdisciplinary focus to a lot of these kinds of communications problems, much more so than seems to be possible to do within a given company. The kinds of things we are looking at are with the university community within our WAMIS program seem to be 5 years beyond a lot of what is going on in industry. I have given a couple of examples here. One being our investment in high bandwidth, high frequency radios that are based on CMOS. Most of what industry is working on today is mimic technology firmly committed to Gallium Arsenide capabilities, as the basis of building those radios. We know that's the way industry is going, so what we are investing in, by some wild and crazy people in the university, is advanced CMOS technology. They believe they will be able to build very high speed technology that will aid communicators based on this kind of technology. No one in industry seems to be looking at that. So sometimes universities can look at very far-out things. There might be an opportunity in terms of teaming with some of the defense contractors. This whole issue of adaptability is of great interest and I think, again, there are a lot of interdisciplinary disciplinary opportunities there. What is the best scheme for dealing with bit error rates within our communication system? You can do things at the signal processing level, you can do things at the network protocol level. Where is the best place to put your emphasis, to come up with effective, optimal technologies.

So that gives you a little appreciation for the kinds of things that we have done in TRP. We are putting together topics for the 3rd round of TRP. This will probably be something in the range of 600 million dollars. While there is no way that the only topic on that solicitation will be wireless communications, the chances are excellent that some aspect of wireless will appear in that solicitation.

20

ESL'S Electronic Warfare Capability for the Cellular Industry

Bruce Cole
ESL
495 Java Drive
Sunneyvale, CA 94089

Good evening Mr. Moderator, members of the Bradley Department of Electrical Engineering and distinguished guests. Thank you for inviting me to share our experiences at ESL, a TRW company, with you today. As one of ESL's new venture managers, I am part of the team responsible for implementing the company's defense conversion strategy. ESL has a unique and successful approach to defense conversion. It's unique because we've developed an internal process by which our employees can bring forward ideas, and turn them into viable businesses. It's successful because we have launched six new ventures in the last eight months.

ESL is a subsidiary of TRW Inc. TRW is a diversified company, focusing on automotive, information processing, and space and defense, with 1993 sales of $7.9 billion. ESL's sales represented about five percent of that total, nearly all of that was in defense.

Since ESL's founding 30 years ago, our focus had been on developing defense electronic systems for the United States government in support of national defense and its goals. However, the fall of the Berlin Wall in 1989 signaled the end of the Cold War and certain change for our industry.

At that time, we began to rethink our future, even though ESL was financially healthy. Today, ESL employs approximately 2,300 people, with headquarters in Sunnyvale, California, and field offices in Sacramento, California and Reston, Virginia.

ESL's approach to defense conversion has grown from a series of trials and errors. Our key success factors include: (1) a commitment from senior management; (2) a willingness to apply substantial resources toward achieving the stated goals; and (3) substantial growth in the industries we have pursued.

Out of concern for the company's long-term survival, ESL President Arthur Money established a program in early 1992 to reinvent ESL's future -- focusing on survival in the year 2020. We began our "2020 Program" with a series of workshops to transform the thinking of grass-roots-level employees. We brought in futurists and strategic consultants to help identify core skills and how those might be applied to new markets. By the end of 1992, six viable ideas for new businesses had emerged.

During 1993, we created a process and infrastructure for evaluating and developing new business ideas. First we assessed our core competencies that map most directly with some of the major market trends today. We hired consultants to advise and train us in commercial business and marketing. We opened an internal incubator center. Lastly, we set a strategic goal for the year 2000: to have 25 percent of ESL's sales in non-defense business.

By the end of 1993, we had evaluated 18 ideas and launched six internal ventures. Examples of some of the ventures launched are:

*TRW PhonePrint^{TM} which blocks fraudulent cellular phone calls, is based on our core competency in radio frequency fingerprinting. The beta version of this system now is being tested in Los Angeles and New York. We conceived, developed, and placed this product in field test in less than 10 months.

* Travinfo is an advanced traveler information system that we are developing under contract for the Bay Area Metropolitan Transportation Commission. The system is intended to provide multi-model transportation information to San Francisco Bay Area commuters. Capabilities include access and display of highway advisories and messaging signs inside the consumer's vehicle. The product is based on our system integration competencies.

* TRW Smart Search^{TM} is a graphical user interface that provides a common, easy-to-use way to retrieve information from multiple on-line databases ranging from the Internet to in-house data repositories. This tool lets users run simultaneous queries without having to worry about data communications paths, passwords and log-in procedures. It can be used on a Personal Digital Assistant, in a vehicle as well as on the desktop.

Those are just three examples of the new products we have introduced so far. ESL has entered three new industries: telecommunications, transportation infrastructure and information processing. Some of the technologies we have considered for commercial products include databases, multimedia development tools, geographic information systems, wireless chipsets and artificial intelligence applications. Over the next five years, we expect to receive revenues from those businesses, retain a significant number of our people, and hire others. It means job retention and job creation.

The government can help to preserve and enhance our success. It's very difficult to create and operate commercial businesses within a defense infrastructure. The primary reason for this is that the rules and regulation involved in government acquisition are very different than those in commercial practices for the same functions, such as security, accounting procedures, materials procurements, contracting practices, and intellectual property rights. It is difficult for us to use the same employees to work under both sets of rules. It's too expensive to duplicate entire departments -- one for commercial and one for defense -- in order to support two very different infrastructures. That is why we support any legislation to help reform and streamline the government acquisition, procurement, and reporting process to make it more compatible with commercial practice. This would not only introduce efficiencies to our defense business, but it would also allow us to use the same employee base for both businesses.

"Conversion" tax credits could help because they would free up immediate funds, at least within the tax year. These would be an expansion of R&D tax credits, specifically targeted for commercialization. Presently, the R&D tax credit is allowed only for an incremental increase in R&D spending over a three-year period. In this time of shrinking defense business, defense contractors are generally spending less on R&D than normal and therefore can't take advantage of this credit. If the formula could be changed so that a certain percentage could be allocated for defense commercialization regardless of previous spending, this could create millions of dollars for conversion. It would create greater incentives for defense companies to expand into commercial markets. I request that you look into what could be done in this area so that defense contractors could be more competitive.

Also, government-sponsored retraining programs in commercial business skills would be valuable. Most current programs focus only on retraining displaced workers. I'm suggesting that programs be available to train people who are still employed, so that they can learn commercial business operations and apply their skills immediately to new markets.

Both of these approaches would allow defense contractors to more quickly transition their work force to be competitive in other markets.

We have learned some valuable, and sometimes painful, lessons:

1. Start the process when your company is healthy, not when you're downsizing. We started at a time when we were stable. Start small and build from resources. Most of our venture teams are made up of less than 20 people, and they operate in a start-up environment. When revenues increase, the work force will expand.

2. You need leadership and support from the top. Without it, people are unwilling to change their paradigms -- their ways of thinking. Senior management needs to be committed to the mission, and must communicate the message.

3. Recognize your core competencies and skills and use them for competitive advantage. Recognize missing skills (such as commercial marketing, in our case) and hire the best from the external world.

4. Be willing to retrain people in basic business skills. Highly technical employees need to learn marketing, finance, and commercial business operations.

5. Focus on market pull -- that is, what the market needs -- and not on technology push -- for example, don't force existing technology into other markets.

6. Use commercial high-technology business plan models to evaluate ventures. We use business plan models taken from successful Silicon Valley start-up companies. We hire local consultants to advise us.

I commend Virginia Polytechnic Institute for holding this very important Symposium. I also extend an invitation to you and all of the participants here today to come to Silicon Valley and visit us at ESL. Thank you for the opportunity to present our unique success story.

21

Hazeltine Corp. Defense Conversion to Commercial Wireless Communication Products

Henry L. Bachman
Hazeltine Corporation
450 East Pulaski Road
Greenlawn, NY 11740-1600

I am going to take the liberty of using some charts that we used when we introduced our new family of wireless communication products at the CTIA show earlier this year. I thought this might add reality to this new emphasis on defense conversion and at the same time I will try to respond to some of the questions that I was asked to consider. Namely, how we were founded, what are our defense skills, how do we apply them to commercial product development, and the benefits they have created.

Hazeltine has been a communications company since 1924. I'll tell you a little more about that in a minute. Our style through this entire period and our strategy has been to leverage technology to provide innovative solutions to satisfy customer needs. We have used technology as a product discriminate. As you will see later, we are a moderately sized company but we compete with the bigger companies and we succeed by the application of technology discriminates. We will apply this same principal as we venture into commercial markets. While it is probably not well known, we have always been a communications company with a tradition for many "firsts". Hazeltine was founded 70 years ago when Professor Alan Hazeltine invented the Neutrodyne Radio. This was the first practical radio to apply the triod vacuum tube amplifier. Hazeltine's business up until WWII involved research and inventions for the broadcast and entertainment industry and the licensing of patents. Automatic volume control was a Hazeltine invention as were many others, including the key inventions that made compatible color TV possible. Hazeltine developed the communications satellite antenna on the first communication satellite, Telstar, launched by Bell Labs. Later, we applied our TV display technology to the development of the video-display terminal, which became the replacement for the teletype, as the means for interfacing with computers. We built hundreds of thousands of these and in the 70's, at a time when we were mainly in the defense business, the VDT's contributed 50 % of our revenue. This year we have introduced new wireless infrastructure products.

First, just a brief description about our defense technology background. Since WWII, we have been a major defense supplier. We were asked to serve the needs of the military with the outbreak of the war. We were given an assignment to develop electronic identification systems that would prevent the newly developed radar-directed anti-aircraft guns from shooting down friendly airplanes. We supplied the identification equipment (IFF) to the allies during WWII; this continues today as a major Hazeltine product. In the 60's, airplanes could carry weapons that could shoot down airplanes beyond the line of sight, but we could not use them, because the rules of engagement said that you had to see the target. At that time, electronic-identification interrogating equipment could not be installed on airplanes. It was not until Hazeltine applied an innovative antenna solution that airborne interrogators became practical and we sold thousands of them. Today we continue to use innovative technology to newer phased-array antennas for airborne IFF systems.

In communications our emphasis has been to satisfy the defense needs for survivable and jam-resistant communications using adaptive antenna systems. The first was probably 30 years ago, developed for radar and later for interference suppression for communications. We built the first GPS receiver in the late 60's, the first spread spectrum receiver ever built, and we have been doing spread spectrum work for 30 years. For communication infrastructure we developed a packet radio system, a self configuring packet radio network which has found many applications. So we have a long heritage in communications technology.

Another very essential element to the defense-conversion process, is the need to be a world-class manufacturer. We were ISO certified in 1993, one of the first defense contractors to receive that certification. That was no surprise. In 1987, we set out to become more competitive in the defense industry and established a TQM program modeled after Motorola's. Since then, we have made great strides in defect reduction and productivity improvement which I won't get into, but are obviously very essential to being competitive in the commercial as well as the defense environment today. By so doing, today we have a manufacturing system that can support both military and commercial applications, so called, to satisfy the need to maintain the defense industrial base as both a supplier for defense and commercial products. In the past, when we were building video display terminals we could not do that. We had to completely separate our military and commercial businesses. We used different buildings, with different people, different systems, and different management. You don't have to do that any more.

Hazeltine, as I said, is a modest sized company, with revenues of about $100 million per year. We have, today, about 70 million dollars backlog in communication R&D. Such programs include a software programmable, multi-function, multi-mode radio and an advanced digital radio intelligent antenna system. Consequently, we see additional opportunity to leverage advanced communication technology into the commercial world, as we have in the past. In selecting the area for entering commercial markets again, we focused on our extensive capabilities in communications. We looked at wireless communications as offering the greatest opportunities. We selected markets similar to those that we were familiar with. We were familiar with operating with a DoD- type of customer. We looked for commercial customers who had some of the same attributes; those that understood complex systems, were technically astute, and relatively few in number. We picked industrial infrastructure products rather than consumer terminal products. I believe that whenever you try to increase your scope of activity, the secret to being successful, is to widen your sphere of activity around what you are already familiar with, rather that jumping off into some new area. That's how we selected our new product/market machine. We then find out what the customer needs and see if we can "scratch his itch". Here we owe a special vote of thanks to Ted Rappaport and Jeff Reed who assisted in selecting product areas. We invested in a new antenna and communication systems test facility. We have over 30 people working and we are applying what we have learned from our TQM program. We introduced a series of wireless infrastructure antenna systems. We also intend to apply our digital- signal- processing radio technology to the next generation of base station radio projects. We introduced this March, at CTIA, a conventional-type cellular antenna. As I will point out in a minute, they are not so conventional. But the objective was to introduce ourselves in the market place, to become known as being in the business quickly, while at the same time developing a more complex intelligent antenna system. This will be a switched multibeam antenna system which provides more gain for better coverage and service to low power portables and reduced interference for better quality or more capacity. We have applied what we know about how to design and build things for the military. Antennas with very low intermodulation distortion and very high reliability and durability that are unobtrusive. We will have another demonstration system in the field in the fall. We intend to demonstrate that we can satisfy the needs as we learned them from our potential customers.

22

MELTING SWORDS INTO PLOWSHARES
Jerome C. Woodard
Signal Science, Incorporated
2985 Kifer Road
Santa Clara, California
2 June 1994

INITIAL COMMENTS

The speakers thusfar have been describing the defense conversion activities in large companies with thousands of employees and annual sales of hundreds of millions of dollars. I will talk about these activities in a small company composed of only thirty employees with annual sales of five million dollars. To accomplish this, I will respond to the seven questions that Dr. Jeffrey Reed sent to me by splitting these questions into two groups. I will start my talk with brief replies to the first group of questions that inquire about the defense characteristics of my company. The second part of my talk addresses the remaining group of questions by providing a philosophical view of defense conversion strategies.

Before I proceed any further, I wish to thank Dr. Reed for inviting me to take part in this panel discussion. I had the good fortune of working with Jeff some years ago when he was a Signal Science employee. He made many fine contributions to the success of Signal Science before he returned to academia. So, Signal Science's loss is Virginia Tech's gain. Jeff is a talented and admirable man, and you are fortunate to have him.

SIGNAL SCIENCE CHARACTERISTICS

Now, the first group of Jeff's questions inquire about my company's basic defense skills, about the wireless products that my company employed in defense, and about the way in which these defense skills and products were employed to transition into the commercial marketplace.

DEFENSE SKILLS. Signal Science was founded in 1974 to provide communications and radar signal processing services to the Department of Defense (DoD) and other US defense-related agencies. In 1978, Signal Science's business activities expanded to include development and testing of signal processing prototype systems. In 1984, Signal Science also began to provide system engineering & technical assistance (SETA) services to US government signal processing program offices.

WIRELESS DEFENSE PRODUCTS. Until 1992, all Signal Science revenues were obtained from sales to approximately one-dozen US military and government agencies. But during 1992, several semiconductor companies who were familiar with Signal Science's government work approached us and asked for assistance in testing spread spectrum chips that they were developing for wireless LAN applications. As these efforts continued, the semiconductor companies became impressed with the Signal Science CAD program that we had developed in our government work to emulate various communication systems. They recommended us as a system testing service provider to several cellular telephone companies. Then in 1993, our non-government work expanded further when we formed a strategic partnership with another US company to provide base station products to the cellular marketplace.

TRANSITION PRODUCTS. The Signal Science MODEM emulation and design skills that we had used for our government customers were applied directly to product development opportunities in the non-government marketplace. Our telecommunications system integration and testing skills also were directly applicable to the non-government market.

PHILOSOPHY OF DEFENSE CONVERSION

I now wish to move on to the second group of questions that Jeff has asked. These questions inquire about mistakes that companies might make in defense conversion, about the relationships among government, industry, and academia in implementing this conversion, about opportunities in the defense marketplace for commercial firms, and about the culture changes needed for a company to convert from defense to commercial work. This portion of my talk will focus on the philosophy of defense conversion rather than on a specific "blueprint" to follow during the conversion effort. I believe that a grasp of the fundamental parameters affecting defense conversion should be more useful to defense companies than is a simplistic set of directions.

MISTAKES IN DEFENSE CONVERSION To start with, I note that we humans have a long history of attempting to convert war equipment into peaceful uses. In the Old Testament if the Bible [*ISAIAH I,18*], it is written: "They shall beat their swords into plowshares, and their spears into pruning hooks." I do not wish to dispute Gospel, but I suspect that a bent sword makes only a mediocre plowshare. However, the sword does contain the proper material for a competent plowshare, but we must melt the sword down and start over with a new foundry casting to make the best plowshare. Likewise, I believe that most defense companies cannot be bent into commercial companies. They first must be melted down, and then recast into the proper form.

The major obstacle to conversion that must be overcome in defense companies is the military weapons mindset (also called the "defense culture") that mandates super-high performance, even if the price of the product is high and the speed to market is glacial. This military weapons mindset causes the companies who manufacture products for DoD to assume that technological superiority will always win the day, even if it takes five years to create the first generation of the product. This plodding approach is necessary to produce military equipment that must work reliably under dreadful conditions to avoid the loss of lives in battle. But, a five year development schedule for "alpha testing" of a new product is a recipe that guarantees failure in the dynamic commercial marketplace. So, defense companies must change their mindsets dramatically if they wish to thrive in non-defense enterprises. I will talk more about this topic of culture later.

Defense companies must recognize that the irreversible decay of the US defense budget has created a crisis for them. I offer three comments regarding the behavior of organizations (and of humans in general) during times of crisis that I have extracted from a fascinating book by Daryl R. Conner entitled *Managing At The Speed of Change* (Villard Books, NY 1993).

(a) An organization must realize that a dramatically altered environment creates a crisis in which both danger and opportunity can be found. Connor's text emphasizes this idea by noting that the Chinese symbol for crisis (see Figure) is composed of two characters -- the top character represents potential danger, and the lower one represents hidden opportunity.

(b) An organization, especially a rigidly structured one, may not recognize that a crisis exists and therefore would not realize that sweeping changes must be made to survive. To illustrate this situation, Connor tells the following story.

危機

"When the British army first faced the Gatling gun (the ancestor of the modern machine gun) in battle, they were stunned by the enormous number of casualties that it inflicted. In the nineteenth century, the British troops still marched onto the battlefield in long lines to face the enemy. Their bright uniforms and antiquated tactics cost them 500 men in a manner of minutes against the rapid-fire Gatling gun. The last battlefield communiqué from the field commander reveals the way in which preconceived notions [failure to recognize a crisis] can block our creative problem solving. The communiqué asked, 'How are we going to get another 500 men?' For the field commander, there was only one way to march into battle. Replacements were his only concern; alternative maneuvers were not considered."

c) An organization may sense that a crisis exists but may refuse to acknowledge the need for change until the situation becomes so desperate that survival is unlikely. Once again, I borrow a story from Conner's text.

"On a July evening in 1988, a disastrous explosion and fire occurred on an oil drilling platform in the North Sea off the coast of Scotland. 166 crew members and 2 rescuers lost their lives. One of the 63 crew members who survived was a superintendent on the rig. He told of being awakened by the explosion and alarms. He said that he ran from his quarters to the platform edge and jumped 15 stories from the platform to the water. Because of the water's temperature, he knew that he could live a maximum of only 20 minutes if he were not rescued. Also, oil had surfaced and ignited. Yet he jumped 150 feet in the middle of the night down into an ocean of burning oil and debris. When asked why he took that potentially fatal leap, he said, 'It was either jump or fry.' He chose possible death over certain death."

The pity of it all is that the superintendent could have lessened his risk of death if he had acted sooner. If he would have gotten out of bed and investigated when the initial confusion occurred, he would have had time to put on a life jacket, to work his way down to the first story of the rig, and to lower himself into the water from there. But in his sleepy state of mind, he did not sense trouble and react immediately when the first indications of the problem reached him.

The punch line of these three stories is that a defense company's failure to recognize that a crisis exists and the company's related failure to react forcefully and rapidly by making massive changes are the biggest mistakes that the company can make while attempting to convert to commercial work.

RELATIONSHIPS. I am very troubled about the proclamations in the media regarding the "proper roles" of government and academia in facilitating conversion of the defense industry. The statements that I have seen all miss the mark by wide margins. In my opinion, these statements all fail to offer much hope for success in defense conversion efforts, because they are based on a total lack of understanding of the fundamental duties of industry, of government, and of academia. I wish to reorient these stalled and futile discussions by providing explicit definitions of the interrelationships and duties of the three entities. Specifically, I hold the following views that I have highlighted to indicate their importance.

(a) The single duty of industry is to make a profit. All else is optional.

The audience must realize that "profit" is not the excess money that companies gouge out of their customers. Instead, profit is a word that is used to identify three additional types of expenses:

> (1) Income taxes (i.e., revenues paid to local, state, and federal governments),
> (2) Rent paid on long-term operating funds (i.e., interest and dividends), and
> (3) Savings for a rainy day (i.e., retained earnings).

If industry can pay these three expenses, then the government will have money to operate, the financial institutions will have money to encourage job formation, and the employees will have secure jobs, both during boom times and during recessions (and therefore can continue to pay the college tuitions of their children). So, profit is not a dirty word. Profit is the engine that pulls a society out of its primitive origins and permits it to flourish. Consequently, generating profit is the only duty of industry in producing a thriving society.

(b) The three duties of the government are to provide for security against outside threats, to assure internal order, and to provide an honest currency. All else is optional.

If these duties are satisfied, then the population can function without fear of international pirates, without fear of riots and disruptions of their activities, and without fear of financial ruin due to erratic cost changes. These three actions create a stable society that can reliably predict the future and can act confidently based on these predictions.

(c) The two duties of academia are to encourage the citizens to learn constantly throughout their lives and to provide a safe haven for eccentrics. All else is optional.

With regard to the first duty, I claim that academia does not "teach" -- the best it can do is to produce an environment where people want to learn. If all members of a society constantly learn, they will not oppose change as a threat to their survival but will welcome change as an opportunity for further personal fulfillment. The encouragement of learning is a crucial duty for academia, because we humans have an innate fear of the unknown. The experience gained in the learning process builds our confidence and causes this <u>fear</u> of the unknown to be transformed into an <u>appetite</u> for the unknown. With such an appetite, all good things become possible.

With regard to academia's other duty of providing a safe haven for eccentrics, I assert that every society needs eccentrics to provide the creative sparks that swirl around and ignite the fires of discovery. The new understandings that come from these fires are the keys that assure the society's survival in our constantly shifting world. But since the early 1970s, academia has become derelict in its duty to provide "ivory towers" that protect and nurture these original thinkers. For example, tenure was one of the tools used to build these ivory towers wherein original but unpopular thoughts could be expressed without fear of retaliation. Now at best, tenure appears to be an employee benefit along with paid vacations and dental plans. At worst, tenure has been prostituted to become a bludgeon that assures herd behavior among the faculty and staff.

I am outraged that many campuses have focused entirely on enforcing politically correct speech rather than on attracting and nurturing original thinkers. The faculty and staff at these campuses have imposed their own shabby, naive preferences on their colleagues, but their "victory" has been obtained at the expense of poisoning the learning environment which now has expired.

Where have yesterday's eccentrics gone? Where are those astonishing mavericks who had questioned the idea that the stars revolved around the earth, who had ridiculed the belief that witches could cast spells, who had loitered around train stations and compiled the hat sizes of the passengers to create the concept of statistics, and who had inquired as to how migrating wild fowl could navigate at night and in fog?

Do any of these precious diamonds still exist? Alas, I cannot find them anywhere. But, they remain at the top of my list of missing persons. My most optimistic theory is that they still exist but have disguised themselves to avoid detection and punishment. We all have been made poorer because of their fear and silence.

So, I am sad to say that the two duties of academia have now been reduced to only one. The process of fostering a learning environment for all citizens is the one surviving duty of academia to assure a thriving society. The benefit of the other duty is lost to us.

(d) There are several optional "duties" that have been undertaken by the government, by academia, and by industry which are wrong-headed and damaging to the progress of our society. These activities must cease, and the participants must return to "neutral corners" where they can operate from their own area of expertise.

Specifically, I claim that:

(1) The government should not use its own employees to perform independent research and thereby compete with academia,

(2) Academia should not try to develop patents for its own financial benefit and thereby compete with industry, and

(3) Industry should not react to this ambiguous situation in a resentful way by minimizing its beneficial contributions to the government and academia. Instead, industry should collaborate with the government and academia as an equal partner and should proudly generate profits that nourish its partners and provides them with operating funds.

OPPORTUNITIES. Another of Jeff's questions inquired as to whether there might be opportunities for commercial firms to sell their products in the defense marketplace. The defense labs that I visit have recently changed their acquisition strategy for prototype systems and for engineering demonstration systems. In the future, they intend to use their own employees to develop these systems in-house rather than to obtain these systems as completed items from the defense industry. The success of this entire concept is predicated on the expectation of assembling these systems from low cost "components off-the-shelf" that are purchased from the commercial sector.

If Dr. William Perry as Secretary of Defense is successful in sweeping out the cumbersome defense procurement procedures so that commercial products are declared to be universally acceptable in DoD, then the defense marketplace will become extremely attractive to commercial companies.

CULTURE. My final comments reiterate the culture changes that defense companies must make to convert to non-defense work. Please underrstand that the term "corporate culture" identifies the complex collection of beliefs and behaviors that a company's employees have developed over time. Culture acts as a stabilizing influence in a company (or in a society). As a result, culture motivates the employees to take actions that justify and reinforce existing business operations. In this regard, we should realize that the expression "culture" in a defense company is synonymous with the word "bureaucracy." An extensive bureaucracy affects a defense company in the same way that extensive hardening of the arteries affects a human being.

I believe that there must be a rebirth of culture in any company which is undertaking the defense conversion task. This rebirth can only be accomplished by "slimming down" the company (i.e., substantially reducing the number of employees) and by "flattening" the company's organization chart (i.e., discarding the traditional functional pyramid structure). So, to convert a company from defense work to commercial work, one should first demolish the company and then build a new company from the salvageable material in the rubble. If this demolition is not accomplished and the existing defense culture is not totally destroyed, then the company's bureaucracy will blindly attempt to assure its own survival by "strangling the baby."

23

Challenges in Defense Conversion

Joe Kennedy
ERA
1595 Springhill Road
Vienna, VA 22182-2235

It's difficult to be the last speaker. So many of the important things have already been said. It is easy to say I agree with Bruce Cole of ESI on one issue or disagree with Henry Bachman of Hazeltine on another, but I will try to provide my view on the issues which have been discussed. ERA is a company of about 300 people and is headquartered in Vienna, Virginia. We are a subsidiary of E-Systems. E-Systems is a $2 billion dollar corporation, exclusively a defense electronics firm. From a size standpoint, we are somewhere in between a small company like Signal Science and a larger company like ESL. We made the decision to diversify in about 1990, in a very similar fashion to what Bruce described. We went through the thinking process very much like what Henry described and that was: "How do we take corporate competencies and apply them to new markets?" So I think the ERA's paradigm is pretty much the same as ESL.

Let me tell you very specifically about some technology we have applied to the non-DoD markets, and some of the specific steps that we took along the way. We have all heard a great deal about digital signal processing (DSP) algorithms in the last day or so. They have been billed as the overwhelming answer to most of the difficult technical questions which this industry is posing, and I tend to agree with that position. ERA started out dealing with very sophisticated DSP-based equipment in the mid '70's. We have been predominately an electronic surveillance and reconnaissance company for the DoD, so our job has been to build very sophisticated receiving equipment.

We started out working in the HF band. If any of you have worked in the HF band, you know it's a very challenging band from a dynamic range and signal density standpoint. It has one advantage, though, there is not many MHz of spectrum down there. ERA actually started out building wideband, fully digital systems to process the signals in the HF band in DSP's using software-based algorithms. We had real "fire breathing" computer equipment like PDP11's when we first got started. Our first systems digitized 750 kHz of bandwidth with about 55 dB of spurious free dynamic range. We have brought that technology from the mid '80's now into the mid '90's where we digitize 10 to 15 MHz of instantaneous bandwidth with about 80 dB of spurious free dynamic range. We are currently on our 5th generation of wideband system.

When I say wideband, what I mean is that we take a piece of RF spectrum, we move it to baseband, we digitize it at very high dynamic range and we do the filtering, heterodyning, demodulation, demultiplexing, and decoding all in software or in DSP-based ASICs. We took this approach to building receivers originally because we were looking for the maximum performance possible. It was the only way to process the HF band effectively. So the motivation was absolutely the epitome of the DoD, which is to build the best "mouse trap" you can to capture the last possible dB of performance. As we went along, we were able to provide high dynamic range, more instantaneous bandwidth, and more signal processing capabilities to support the DoD.

So for about a decade, the key advantage of wideband digital signal processing communications systems was performance. Then in about 1993, another advantage came about. These receiver architectures became cost competitive with their analog equivalents. We found ourselves with an opportunity that was created by the ever decreasing cost of a "MIP" of processing power and a VISI-based DSP function. We now believe that we can build receiving equipment cheaper than the analog equivalents and quite often provide a performance edge.

Given this technology and our desire to diversify, we took a slightly different tact than the other companies on the panel for our first defense conversion initiative. We identified 2 markets areas that we wanted to explore. Our first endeavor was a very small step from our traditional business area. Given that we predominately sell to DoD, we attempted to do business with another part of the federal government. About that time the ISTEA legislation was adopted by Congress and the Intelligent Vehicle Highway System program came into being. We began to try to identify places where we could apply our technology in that market. Last year 10% of our bookings were from the Department of Transportation (DOT). We considered ourselves successful in making the transition to a non-DoD market. It was a small step because unlike the DoD markets, cost played an important role, but more in the long term than the short term. In the short term it was still technology innovation that was the critical factor in our competitiveness

The second opportunity we have identified is the cellular/PCS markets. We understand how to build very sophisticated receivers. We believe we have a very innovative approach in the market through our wideband DSP-based technology. Like Hazeltine, we are not well qualified for small equipment, high volume production, so we are going to pursue the infrastructure side of the industry where the products in the market are more like what we have produced for the DoD. We are well qualified to build hundreds or thousands of rack mountable assemblies. We now have a set of products and cooperative projects including beta tests organized with service providers in the industry for base station equipment and adaptive antenna arrays.

With respect to the lessons learned in our defense to non-defense market conversion attempts, I think there are three key items. First, the business paradigm is totally different. When you do business with the DoD, you are paid to do R&D,

you are paid to do production and you are paid to provide support for the equipment. In this model, your up-front investment requirements are not large, but neither are your profit margins. As you know, profitability is limited through the government contracting mechanisms. In the commercial world you must fund your own R&D and potentially your product support, but your profit margins are much higher on the equipment you produce. So there is a major shift a company must make in its business model in order to succeed in the non-DoD markets. Up-front investment is required, with its associated risks. But the potential rewards downstream are much greater.

The second key item you have to look at is motivation. In the DoD, a more capable "widget", 99 times out of 100, is the motivation. Every dB we could extract on the performance side of our system designs was worth a great deal of money. The over-arching motivation in the commercial industry is dollars. Businesses are driven to make profits and an equipment vendor must find ways to save investment dollars or increase revenues to sell products. In the wireless industry, the desire is to increase capacity or coverage, reduce infrastructure cost and reduce operating cost. These are the motivators.

The last key item has already been discussed in detail. It is timing. We have been through product life cycles in the DoD where the time span from program conception to operational capability was 10 years. In our IVHS endeavors, from the signature of the contract to the delivery of the equipment has been 10 months. These shorter time scales are reflective of the norm in non-DoD markets. The ability to change the business model, invest up front and take the risk, and move quickly are key to successful transition.

We have had experiences similar to many of the ones described by the panel members. I'll highlight a few. From an organizational standpoint, we have not split off our non DoD work to separate facilities, but we are a fairly small organization. We do tend to segregate the group to create a little bit of a different environment. I concur with Bruce Cole, it is imperative to receive the top management commitment early for there to be any chance for success.

"Competitive Issues in PCS"

by
George Morgan
Center for Wireless Telecommunications
Professor of Finance
Virginia Tech

The first panel discussion of the Conference focused on issues associated with the "emerging technologies" auction of bandwidth for personal communications service. Rather than discuss technical aspects of PCS, the panelists described several important business and regulatory issues surrounding PCS. Two panelists discussed the design of the process for allocating licenses. Those discussions highlighted the importance of several variables in assuring a fair and efficient allocation of PCS licenses -- as well as meeting the objective of maximizing the revenue received by the federal government. The other two panelists discussed some alternative competitive mechanisms for offering service to customers. In one sense, this latter topic served to reduce concerns about the "optimal" design of the auction process because unlicensed PCS and ESMR were described as viable and in some cases even preferable means of offering service. If the auctions were not conducted in the appropriate manners, losers (and winners for that matter) have an opportunity to use other vehicles. The speakers agreed that the outlook for PCS is not bleak, but PCS had a better chance of success with an appropriately designed auction, fewer delays in licensing, and adequate safeguards against monopoly or duopoly.

Before the auctions take place, much thought and consideration must be given to the rules for allocating licenses -- who is eligible, what restrictions will apply, what requirements must be met over the life of the license.

The main issues considered by the regulators are spectrum management, universal access, inclusion of women and minority owned businesses, and avoidance of domination by multi-billion dollar businesses. The motivations and rationale shaping the FCC's decisions related to PCS were explained by Mr. Byron Marchant, Chief Legal Advisor to Commissioner Barrett. Mr. Marchant noted that the licensing authority for PCS came from the Omnibus Budget Act and thus revenue production is certainly a driving force in design of the process. However, he expressed his primary criterion for any regulatory decisions on PCS: they must "essentially enhance the competitive infrastructure of the communications market." In designing a band plan, for example, competition between at least three alternatives to cellular providers requires allocating sufficient bandwidth to support each company's profitability and therefore its ability to attract capital. This is quite a departure from the cellular phone licensing process where duopoly power was instituted in part for the same reasons of promoting investment in infrastructure.

These concerns dictate "promoting a licensed block of significant size that would be in the basic trading area that would clearly have alternative choices for competitive niche services..." Concerns about time to market, synergies, cost, and coverage also entered the creation of the band plan. (Since the conference, the FCC announced a revised band plan that increased the number of 30 MHz licenses and moved the unlicensed band for PCS.) Care must be taken to allow synergies to develop at the time of the auction which will in turn promote ubiquitous service that the consumer will find convenient to use.

Following determination of the "macro" rules of the licensing process, careful consideration must be given to the implementation of auctions. Professor Charlie Plott, an expert on the theory and practice of auctions from Cal Tech, pointed out that seemingly innocuous, technical details of auctions can have significant effects on who wins and who loses. There are basic economic principles that drive auction behavior and these are observable in experiments with human subjects and are easily replicated.

Auctions force people to "put their money where their mouth is." The forces of supply and demand will make sure that the distribution of licenses is systematic as well as efficient, **if** the rules of the auction are set up correctly. PCS auctions are quite different from the standard controlled auction experiments that have been traditionally examined. With simultaneous, multiple round auctions, complexity is guaranteed. Professor Plott examined two of the many proposals for auction rules.

First, consider the format where a single sealed bid for a nationwide license occurs and is followed by a sequence of auctions of licenses starting from highest to lowest value. If the sum of the individually auctioned licenses exceeds the sealed bid, then the licenses are awarded individually. Otherwise the highest sealed bidder wins. If the objective is to award the licenses to the buyer who most values them, this process fails. Those that do win an individual item in the auction now have a stake in it and therefore have an incentive to bid greater than the sealed bid. Those bidders feel that they may end up with something they don't want, but at least they don't lose something they do want. In this situation, smaller entities will be out bid and knowing this phenomenon will occur, the sealed bidders will over bid in order to win.

A second format would open all markets simultaneously. Bidders in this case will be able to put together packages which in the end will result in relative seamless, ubiquitous service. One important component of the rules is called a "release rule" suggested by Professor Plott's group at Cal Tech. A license can be released by a winner of the license and put back in the pool of licenses. Thus if a collaboration or packaging of licenses does not develop as anticipated, the license can be returned and the next highest bidder receives the license. The rule requires that the party releasing the license pay the difference between their bid and the next highest bid. This will inhibit overbidding by parties who want the "option" to release the license later if things don't work out. The release rules are a technical detail of the auction that illustrated Professor Plott's point that "who wins and who loses is a function of the details of how the auction is designed."

The speakers from industry provided interesting insights into some of the alternative to licenses PCS. This is an important area because firms that do not win at the auction will have other means of offering service to their customers and also because there are synergies between the different alternative. Mr. Eng, MCI Wireless Business Development director, described the view that ESMR and PCS are complementary and offer opportunities for those who want to offer personal communications services.

MCI desires to offer particular wireless services to its customers existed prior to the opportunity to acquire NEXTEL. By defining the service rather than the medium for providing the service, MCI sees ESMR as just another way to serve their customers in conjunction with PCS. Mr. Eng sees two uses for PCS. One is the extension of service to a neighborhood or regional coverage where consumers would have both a wireline and wireless/cordless access. Second, is the replacement of the local loop.

One of the most important markets for personal communications is the extension of the cordless phone to a broader area. The area covered does not have to be national in order to provide valuable service to consumers. In fact, service within 100 miles of home could be a very valuable service. Without waiting for PCS auctions and awards of licenses, ESMR licensees can offer this service. Given the concerns expressed by many prospective PCS providers about the delays in PCS licensing and the "head start" that cellular companies have with their 40% annual subscribership growth rates, ESMR is an extremely attractive strategy.

Furthermore, the acquisition of NEXTEL gave MCI a nationwide footprint that is not as readily available with PCS. (The FCC did not rule in favor of MCI's request for auctioning of a national license for PCS.) One concept for enhanced footprint is "terminal mobility." A Smart Card could be carried by the customer and inserted into whatever terminal equipment was available (leased, borrowed, etc.). The customer would be billed for the transmission from wherever it originates, whether in Europe, in a rental car, or in the backyard. Furthermore, the customer's interaction with customer representatives will also be ubiquitous thus avoiding customer dissatisfaction with being "shuffled around" when outside their home area. This synergy in customer service and billing can be an important dimension to a consumer oriented, "personal" service.

Another extremely important aspect of ESMR technology is its compatibility. For some time, MCI has believed that its implementation of PCS would be based on GSM technology. In fact, ESMR is compatible with GSM from the base station controller back to the switches. The over-the-air interface is different, but with dual mode handsets becoming economically viable, the air interface is not a critical issue.

Nonetheless, MCI sees a need for acquiring licenses for PCS spectrum. The strategy is one of supplementing their ESMR capability where the goal is to replace the wired local loop. Especially in markets with high demand for service, MCI will need additional bandwidth (as will cellular companies who will be restricted to buying the 10 MHz licenses in their home markets). Those markets will likely be the top urban markets.

Another alternative to using licensed PCS is to use the unlicensed band. Mr. Lovette, a Principal Scientist for Apple Computer in Cupertino, described why unlicensed spectrum is vital to the competitive and national deployment issues surrounding PCS. The unlicensed band will be a natural haven for those who do not win at the auction. The smaller cell size of say 15 meters will mean serving niche markets but at a lower cost because of the absence of need for 8000 watt base stations.

Mr. Lovette also noted the role of unlicensed spectrum in the development of the National Information Infrastructure. The media glamorized NII but has little appreciation for the costs involved when every school room in the country will be on-line with the capability to see volcanoes erupting in real time. It is tempting to assert that fiber optics will provide the needed bandwidth for such uses, but the cost of running fiber to every school, classroom, desk in the country is astronomical. He believes that this is where unlicensed PCS can compete.

Unlicensed will perform primarily two functions. First, the existence of unlicensed providers will put a cap on what licensed providers can charge for service. Second, unlicensed providers will be feeder conduits to the licensed providers. The analogy is the local area network within a company that is then linked to the Internet. The unlicensed spectrum will generate traffic for the licensed. The degree of synergy and conformity between the two will determine the success of each.

The panelists did an excellent job identifying the issues and the implications of alternative structures for the competition that will ensue post-auction. All the panelists expressed confidence that PCS is a viable market segment despite the lead of the cellular companies and the advantages that the large companies have in bidding for licenses. By careful attention to the details of the allocation/auction process, problems will be avoided. However, even if they aren't avoided in the licensed arena, unlicensed and ESMR will be a viable path for companies to take in providing service to consumers. The best outcome is a well-designed auction process and the creation of symbiotic relationships between unlicensed, ESMR, and licensed PCS.

24

Business and Regulatory Issues in PCS

Byron Marchant
Federal Communication Commission
Washington, DC

I appreciate the chance to speak to you today. It just so happens that when I accepted this invitation to speak months ago, I had expected the Commission to have, by now, made a decision on PCS. Little did I know that I would now be less than a week away from making a decision, again, on PCS. Needless to say, this is a very timely subject. Industry is flocking into Washington and, as is typical of periods before these decisions where no one can come into the Commission, we get calls from the wee hours of the morning until about midnight saying "There's one thing I forgot to tell you..." In fact, in order for me to come down here, I had to schedule a conference call and many other things on this very subject back in Washington because of the extensive interest in it from all corners. I will face these same forces when I get back. It's very comforting to be down here in an academic environment to talk about this for a change.

I have been working with Commissioner Barrett since 1989 in the areas of emerging technologies and wireless communications. He and I both had a vision of making policy and regulatory decisions that would essentially enhance the competitive infrastructure of the communications market as we know it today. It has been his goal, as well as mine, to create an environment where the public, at some point, will view wireless communications as a reliable, price-competitive alternative to the local telephone service that they get from carriers such as a Bell Atlantic company, or from the long distance carrier of their choice.

The future appears to be bright in this area and everybody appears to be interested. It appears that we are going to see conversions of interest in this area, especially among those who have already built an infrastructure in the ground. They can leverage whatever fiber, cable, or copper plant that they have and perhaps acquire a wireless mobile license in personal communication services to serve their customers, their business, or a niche market such as a university. The possibilities are endless and we believe that major developments will occur once we finally develop a PCS licensing structure and establish a more definitive framework for those licenses. It is an exciting time to be in Washington.

I would like to speak to you today about the kind of broad policy areas that the FCC must manage in order to come to closure on a PCS decision by Thursday, June 9th. I will begin by noting that we attempted to make a decision on PCS licensing in the fall of 1993. Commissioner Barrett voted against that decision because he and I felt that sufficient analysis had not been conducted on either the economic potential of license allocations or the actual economics of the license structure. We knew the potential for infrastructure development that could occur. We also knew the potential spur that we could give cellular providers in terms of new competition, and the major evolution that could occur with new industry players not traditionally associated with the telecommunications market. Needless to say, we are happy to see that 67 entities appealed that decision.

We had started the emerging technologies docket about a year and one-half to two years before legislation came from congress last summer. In 1993, Congress became involved and has been a greater participant during the development of PCS licensing procedures. They passed the 1993 Omnibus Budget Act which mandated that the Commission raise revenues by auctioning spectrum and licensing new services as efficiently as possible. In addition the act directed the Commission to make sure that new providers and small entities are not completely eliminated by auctions that may favor those with well-capitalized investment budgets. Congress also directed the Commission to act quickly in bringing PCS to the public.

We have been wrestling with all of these goals, attempting to devise a workable, balanced plan that can be understood by current industry players, who will sink private capital into the licenses, the investment community, and the consumer public. All along, we have kept in mind the goals of providing new services, allowing new competitors, and more importantly, more choice in price competition for consumers.

Now that we appear to be on the verge of making another decision, I would like to go over some of the basic issues that Commissioner Barrett and myself, along with the other Commissioners, and a new chairman, are facing in trying to devise a plan for PCS. A plan that will allow the nation to see new competitive wireless services deployed in a way that their price is compatible with whatever the customer choice, at the time, would be. First of all, there is clearly a spectrum management issue when allocating spectrum for a new service. The Commissioner and I are looking for something that will promote at least three viable, competitive alternatives to the existing cellular infrastructure as well as the emerging ESMR infrastructure that is serving the wireless market today. The GAO and others have studied the cellular market and decided there could be greater competition than exists under the cellular duopoly scheme that was adopted in the early 1980's. This scheme was not the best way to promote competition in wireless telephony. But, at this point a significant amount of capital has been invested in cellular, and a market for mobile telephony has been established. In fact, there are approximately 16 million cellular customers. We are now engaged in an effort to learn from the cellular experience and license a new service with as much broad area coverage as is feasible given our different policy goals.

Our first goal is to ensure that at least three viable entities come into the market with incentives to invest in an infrastructure that will compete for new wireless customers, provide price competition, and supply the kind of roaming and coverage to which we are becoming accustomed. To us, the term "personal communications services" really means a hand set which can be utilized in a lot of areas of the country according to what you are willing to pay, and the kind of convenience you seek. We hope that by having a band plan that will promote three clear competitors to the existing cellular infrastructure and the emerging ESMR infrastructure, there will be more competition and greater customer choice.

The second part of the spectrum issue relates to the spectrum allocation size and market necessary to spur this kind of investment. The Commissioner and I believed, and still believe today, that some major market allocations, on the order of Rand McNalley's MTA's, are needed to attract private capital. We also believe that the MTA allocation is a way to overcome the cellular head-start, get into new domestic markets, and compete for additional wireless telephony and data customers. It is our hope that next week's decision will include an allocation of sizable spectrum and include Major Trading Areas as a part of the allocation scheme.

At the same time, we want to ensure there are opportunities for other entities other than the top ten local exchange carriers, that have several billion dollars in revenue and at least a billion each in cash flow. We also seek to ensure that opportunities do not go only to the top interexchange carriers, who have more revenue and have greater cash flow than the top LEC's. Per the Congressional mandate in the 1993 Budget Act, we want to make sure that our policy structure and our band plan promotes new licensing opportunities for more than just those 12 - 20 players.

We are also interested in promoting a license block of significant size in the Basic Trading Areas. This block could be used for either competitive niche services or as a synergistic block that is applied in a nationwide or regional scheme by larger carriers who seek coverage where they may not win the license at spectrum auction. I hope, and in fact anticipate, that our decision next week will include Basic Trading Areas and an allocation block size which is at least proportionate in size to the 30 MHz MTA that was adopted in the Fall 1993 decision.

The time to market is another issue of concern. I expect MCI to announce, today, that is has invested in NEXTEL. We expect this to result in a mandate for the Commission to act expeditiously to establish licensing of an advanced wireless infrastructure and unleash the potential for investment. Therefore, we are attempting to solve some of the initial time-to-market issues for new players while we develop a band plan. A partial solution would include a spectrum sharing scheme with the encumbered microwave providers in the 2 GHz band as well as in the lower band where we believe there has been a lot of manufacturing prototype support for equipment development. I would hope that next week's band plan would attempt to solve that by allocating spectrum sufficient to support initial sharing. This would minimize the up-front cost of moving encumbered microwave carriers in this personal communications services band.

Beyond that, there are issues of synergies and infrastructure build-out costs. In attacking these two issues, we must also address the necessary power limits and how those power limits effect capital budgets for deploying PCS networks. In 1993, the Commission decided to raise the power limits for the base-stations that would be deployed in a PCS network so that hand-held units could remain relatively light and perhaps dumb. Base-stations and transceivers that would, essentially, route and transmit personal communications traffic, would then possess the "intelligence." Because of the numerous appeals since that decision, we hope next week's decision not only reflects some of the things we learned since the Fall of 1993, but does so in time to effect market decisions by minimizing the necessary capital outlay for deploying PCS networks over a wide area.

Another issue is the kind of service-area coverage a licensee will have to provide in order to keep a license. Two methods of handling coverage requirements have been proposed. One method involves a geographic coverage threshold, requiring the licensee to provide service which covers a certain amount of geography by a certain date and maintain that coverage over the license term. In the other method, population coverage dictates the manner in which a licensee must provide service. Given the precedent of our two-month old decision on narrow-band personal communication services, I expect that we are going to look favorably on the population coverage method. This method would also alleviate some initial burdens of entering markets with existing cellular competitors. Licensees would be allowed to satisfy coverage requirements by serving some of the major markets that will allow them to roll-out service and begin obtaining revenue fairly quickly.

Beyond spectrum management issues, there are issues related to the economics of the mobile telephony market. In Washington, we are actually concerned about two kinds of economics. The first concern stems from the fact that Congress decided to auction the PCS licenses. There is no secret that the licensing authority for these new services came from an Omnibus Budget Bill, and therefore from a side of Congress that was worried about budgetary issues.

Although the Bill, particularly portions dealing with communication issues, was drafted to a great extent by authorization and committee staff in both the House and Senate, it also went through a budget vehicle and was signed by the President in that context.

The concept of auctions is, for the first time in this country, being implemented to license electromagnetic spectrum for communications services. The last thing we want to do is implement auctions in a way that creates more confusion than any other licensing processing that we have in place. Starting this summer, we will face a great challenge in implementing the narrowband PCS auctions. We must apply what we learn from those auctions to the broadband PCS auctions that follow.

Auctioning such a large amount of spectrum is an immense undertaking. Each one of the MTA's represents approximately 50 market areas around the country. Each one of the trading areas represents about 490 market areas around the country. Therefore, you can multiply the number of licenses in each kind of allocation block and determine the magnitude of auctioned spectrum.

Despite the magnitude of the task, we are interested in accommodating both large and small entities during the auctions. We would like to make sure that the auction is consumer-friendly to large entities that have clear synergy strategies for providing a seamless, ubiquitous mobile telephony network. These entities may be existing players with plans to provide service in conjunction with their existing inner exchange or local exchange telephone structure. They may also be cable companies or, perhaps, even utility companies who are wondering whether they should get involved in PCS. In either scenario, we will attempt to allow synergy at the time of auctions that would accommodate those plans.

At the same time, we are wrestling with an auction structure that will permit smaller entities, rural telephone companies, minorities and women-owned businesses to combine efforts and offer services in geographic areas that may not otherwise see the first levels of service deployed by entities with a larger aggregate scheme. This auction structure will have to be developed such that the smaller entities are not always out-bid for service areas by LEC's and other large companies. In addition, the actual license structure will have to allow a mixture of service areas such that, for example, it could be more affordable to offer service in the Roanoke/Blacksburg corridor by a group of rural telephone companies.

Beyond the economics of the auction process, there is the economics of the market itself. Part of our plan is going to make sure that both the final spectrum and market allocation enable joint-venturing by licensees, who can then market to provide a wider range of services. We want to allow any array of services or ventures that will allow licensees to offer the full spectrum of services. These services could range from offering only a walk-around service here on Virginia Tech's campus to offering additional services that would allow a high-speed hand-off between here and Washington, Richmond, or Charlottesville.

We anticipate facing issues in the delivery of these services, some of which are particularly appropriate to mention at Virginia Tech. Both interoperability issues and equal access issues will be raised in new service delivery. I suspect that interoperability issues will be of particular interest in this area. We have been fairly laissez-faire about interoperability in the cellular environment, allowing the industry to develop a standard by itself. This standard was developed in the analog world and, as the analog capacity has been consumed, there has been an effort to either enhance the analog capacity through more advanced services, or, obviously, transition to the digital radio services which are the talk of the time. The problem seems to be that there is no clear agreement on which digital transmission standard to use.

We are especially concerned about this in the PCS area because it is our goal to have competitive, consumer-friendly service. In other words, we hope to see services which are widely-available at costs which consumers are willing to pay. One service should allow a consumer to use one handset, instead of dropping a handset at the border of a certain MTA and picking up another handset, or worrying about whether or not you can roam from you area to another. MCI is attempting to solve the service/price problems by purchasing a certain amount of spectrum and telling customers to join MCI. It is our hope interoperability solutions come from more sources than just MCI and AT&T. Therefore, we are going to be interested in hearing from the industry.

I think it is going to be our concern that we are not just laissez-faire in allowing these issues to drift on for years while manufacturers position themselves to take advantage of one technology over another. How we do this, I am not sure. But I do know that we already have a commercial mobile radio services docket which is an effort to bring regulatory parity to all mobile services. There are some issues in this docket, dealing with interconnection and equal access, which need to be addressed. In addition, I would expect to raise issues in that docket that pertain to interoperability, and what the Commission's role should be, vis a vis standard setting bodies, in making sure that the consumer ultimately wins.

Intertwined in all of this decision-making is politics. After all, it would not be Washington if there were no politics. There are political interests which will, undoubtedly, have some impact on our decisions. For example, the public safety community is concerned about the microwave moving plan which we adopted in March. The plan stated that microwave stations can be fully compensated after 5 years, but can be moved involuntarily if they do not move themselves. I understand they are petitioning to reconsider that decision. A reconsideration may complicate things, depending on the difficulty of sharing with microwave incumbents. This is the PCS deployment problem.

There are also issues pertaining to the involvement of minorities and women in rural telephone companies and small businesses in an auction structure. We must consider whether we adopt bidding credits, installment payments, or any configuration of policy preferences and enhancements that we can adopt. Again, our goal is to make sure that there is a deployment of licenses beyond the top 20 players in the telecommunications industry. We want to allow the kind of innovation which often occurs at the local level, as well as that which may occur between software providers and computer companies.

There are several other points relating to synergy. One is that there is an unlicensed part of PCS, which I have not mentioned. This part is going to involve both local wireless PBX providers and, potentially, local data providers or data providers which have devices that roam the country or a particular region, essentially allowing a user to communicate over a particular band. We hope that there will be synergies between some of the licensed and unlicensed services that service configurations and joint-ventures can offer the customer a wide array of choices.

Finally, there is the mobile satellite industry. A large portion of our PCS allocation decision affects the mobile satellite industry which has planned to deploy services in the late 1990's and the early part of the next century. Because of their long-term planning issues, we have to coordinate with them to make sure that what we do for terrestrial services does not result in inadvertent regression in our movement to provide mobile satellite telephony both in the United States and around the world. There are several major proponents of world-wide satellite systems whose interests we have to balance as well.

I hope to have given you a flavor of some of the things we are attempting in Washington. We are going to move forward here. We think it is imperative that we do so. As we move forward, we invite your input. We will need a significant amount of input from the technical and engineering community to make sure that once decisions are made they can be implemented, networks can be deployed, and customers can be served effectively.

25

ESMR, SMR and the Transition to PCS

Corey Eng
MCI
1133 19th St., NW
Washington, DC 20036

Just to give you a little bit of background about myself. For the past year I have been responsible for auction strategy, partner development, and overseeing some of our regulatory strategy and comments. Prior to that, I spent 2 years with MCI's 800 service in the marketing group. Before that I spent some time with the Walt Disney Company out in Burbank.

What I have been asked to speak about here is how ESMR or SMR Technology might fit into this whole thing called PCS. There are 3 major considerations in terms of MCI's position on PCS. The first consideration is that we wanted a national footprint in the wireless game. In some of our comments to the FCC about a year ago, we asked that we have a consortium and that the FCC consider creating a national license for this consortium. They did not rule in our favor, but we still wanted to have that national footprint. We thought through NEXTEL we could do that. In NEXTEL, with their particular coverage, and with their agreements with other ESMR Technology companies, we thought we could obtain that national footprint.

Another consideration which we thought was very important was speed to market. The cellular companies are not just sitting back there, they are growing very rapidly. Now there are only 2 cellular carriers per territory. So we need to have that speed to market if we want to be competitive.

Probably the last consideration, which is also very important to us, was compatibility of technology. Before the NEXTEL play even came up, we always thought that we wanted PCS, at least for MCI, to be based on the GSM technology. What you will see a little bit later on one of my slides is the ESMR technology and our vision of the GSM, are pretty much compatible. We believe ESMR technology is a good fit with the targeted services in market that MCI wants to go after.

Even before considering the ESMR technology, MCI through their PCS efforts were planning to address three or four targeted services. The first targeted service was mobile service, very similar to cellular service today. For example, you would have a mobile hand-set which would be a second hand-set in addition to the wire line-phone that you had in your house. The next major market that we wanted to go after was replacement service. That is, we actually take the telephone from your house and make you an untethered user. We believe that from our market studies that there is a tremendous business opportunity to establish a regional wireless telephone service, even if it is just within a few miles of home. The third major area that we wanted to go after in PCS was wireless data service. I think one good example of the wireless data service is the little boxes that people like Federal Express are using to track where their packages are. Finally, we are also considering short messaging service. Today, paging is a very big industry, it's thriving. A lot of people need to be reached at all times, no matter where they are. Through PCS we envision that we would be able to address this market as well and in addition to bringing just one way paging, the offering of bringing two-way paging by answering back to a page or sending some sort of text message in response. Through ESMR we are also able to get into another type of service which we weren't initially targeting. That is the dispatch service. Primarily, the dispatch service has to do with taxi's, public safety, and some government services.

There is very much an alignment between what we think ESMR can do and what we wanted PCS to do for us. The primary markets that we wanted to go after in PCS were the business markets. In the business markets you have a lot of people who travel. One of the things we wanted to do there, again, was to have aggregate coverage, and a nationwide footprint. On the residential side, as I mentioned earlier, we believe there is a huge opportunity to provide nationally or even regional coverage. For instance, you could use your cordless telephone or your cordless hand-set within 100 miles of every house. We believe this opportunity is tremendous. The third major step, again, is the mobile telephone service where we would be the second telephone to the wireline telephone that you currently have today.

This is the extent of my engineering knowledge on this subject! Again, when we were looking at PCS, aside from NEXTEL and ESMR technology, we were thinking GSM all the way. One of the advantages that we saw about GSM that it was a proven technology and it is being used throughout Europe and parts of Southeast Asia. One of the things we liked about GSM is the global service compatibility. ESMR technology from the base station controller back to the switches is GSM compatible. With the subscriberships projections that we are forecasting, we believe that the price of dual mode hand-sets will come down so that they can be affordable by consumers and definitely by the business community itself. While the air interface is different for the ESMR technology, we believe the dual-mode hand-set is definitely viable.

The deployment of NEXTEL is up and operational today in LA. It's commercially available, and we plan to have San Diego up probably at the end of June and hopefully San Francisco up and running sometime in the August time frame. I think the bottom line here is that we know that we can't just sit back and wait for the license and the auction process to occur. There is a huge amount of delay that is going to be associated with the deployment of PCS in general. We would like to be in the top 25 markets within the next year and one-half or so that we can offer that competitive alternative to the cellular companies today.

As I get into the competitive issues, I think that what you will see is the importance of speed to market actually has in the technology compatibility. I think that Byron mentioned, the CELCO'S today, there are approximately 16 million users that is representing approximately 6% of the total population of this country. Some of the marketing reports that I have seen recently said that a lot of the CELCO'S are actually increasing their subscribership counts by about 40% a year. Every 2 years they are going to double in their subscribers. If you look at what it is going to take PCS to reach this capacity you are going to see what a head start the CELCO'S already have. With a trench base of 16 million subscribers plus doubling subscribers about every year and one-half to 2 years. It doesn't mean that the CELCO's are necessarily problem free. Because of this tremendous growth, they are starting to think about the capacity problems that are facing them. Because of that they are starting to think about new technologies and using digital communications to increase capacity so that they can accommodate all the subscribers. Unfortunately the technology debate for the CELCO's is pretty much a wide open issue. They are thinking about TDMA, CDMA, a bunch of other technologies, and the companies have their different viewpoints on which technologies. It doesn't seem that all of the different cellular operators are thinking along the same lines. Currently, I think, whoever has a cellular system out here today or a cellular phone out there today, recognizes that they don't have a seamless system today. When I take my cellular phone, which is registered in the Washington DC area and I go to another part of the country, I have to register that phone manually. In some cases, not only do I have to register the phone manually, I have to get a new telephone number and let everyone that is suppose to call me know what my new telephone number is. So obviously that is not a seamless service. Using ESMR, we thought they pretty much had a national footprint with all the roaming agreements that are currently in place. I think they address approximately 96% of the population today. The one nice thing we do like about ESMR technology, is that it is one technology across the board. The system is under construction today. The best part of our system is that it is GSM, and one standard that should allow for seamless roaming.

The main competitive issues facing PCS, at least to the MCI consortium, is enhanced capacity. ESMR, while it does offer the national footprint, we recognize there is just not enough spectrum out there for some of the markets that we want to go after and the service that we want to go after, namely the replacement service. So what we plan on doing, and I think what a lot of the other companies will do, is to include the cellular operators. They will bid on some of the PCS license to enhance some of the spectrum that they currently have. PCS, again, has a very strong vendor support. We have talked with a number of companies who provide equipment from the handsets all the way down to switches, NOKIA, Ericsson, Northern Telecom, they are all just waiting for the auction to go through and the license to be awarded. Right now few companies, in fact I don't think any companies, are willing to sign a contract with them not knowing what PCS license they have won. I think all the vendors are behind it but they are just waiting for the licenses to be awarded. PCS is not without its negatives as well. Going into the PCS game again, the speed to market is very important. We believe that it is going to take anywhere from 18 to 24 months to deploy PCS system in any given market. As Byron mentioned, one of the issues we are concerned about is tower heights and power output. With the ESMR technology the power is much greater so we don't have to have as many cell sights as you would with PCS system. A similar argument can be made for the CELCO's systems. So, the time frame is also very critical to us. Finally, the auctions themselves. I think when I first joined the PCS group we were thinking that the auctions would have occurred sometime last spring. Right now, we are hoping they will occur sometime this year. So if you add 18-24 months to that you can see that is approximately two and one-half years from now that the CELCO's, who are growing at a 40% clip, have a huge head start over everybody.

Our view of how we feel the spectrum is going to be utilized, not only by MCI but other bidders, can be broken down into three different license areas. The 30, the 20, and the 10 MHz. The 30 MHz band, we believe is ideal for the replacement service. To provide competition to the local loop. Our local loop would be wireless. We believe that the 30 MHz is appropriate for the high density markets and therefore MCI would probably bid for the 30 MHz, for instance, in New York, Los Angeles, and San Francisco. MCI is very supportive of three, 30 MHz license, up for auction. I'm not going to speak for Byron, but we are hoping that occurs. What we are hearing is that there is a possibility that they will set aside 20 MHz in addition to one of the 10 MHz license that could be combined and offered into a 30 MHz license and we are hoping that would actually occur. On the 10 MHz license we believe that not only will the 30 MHz license winners bid on them, we believe the ESMR license holders will bid on them and the incumbent CELCO's. This strategy will be just to increase the amount of spectrum they have available so that it can serve the growing market. We also believe the 10 MHz license is very appropriate for those local operators who want to be nothing more than a regional niche play. Again, our marketing forecast shows there is a tremendous opportunity for an operator to go in there and replace the cordless telephone, replace the land-line telephone and offer regional mobile services.

MCI has always thought that we needed to be a national player. Probably one of the biggest reasons that we needed to be a national player was that we wanted to prepare MCI for the total end-to-end competition that is currently going to be coming. There are a lot of bills in Senate and Congress right now, the Brooks/Dingle Bill, etc. that have to do with local competition and letting the local RBOC's get into long distance service as well. We definitely feel that competition is going to be coming for end-to-end services. The characteristics that we are looking for in providing nationwide services is to give the customers and the end users terminal mobility as well as personal mobility. The notion of terminal mobility would mean that you would have what we call a Smart Card which is about the size of a credit card that would slide into the hand set. What that allows you to do is when you are traveling and you don't want to carry your telephone with you, you could just slide your Smart Card into some body's hand set and use that telephone. Even if you were to go to another part of the world, say England or France, you could take this GSM Smart Card, when you rent your car and mobile telephone, the card slides into that telephone and you have service. Even though their GSM system is on an entirely different frequency, the billing structure and billing rating service would all be the same to you.

As far as the same nationwide roaming service again, what we think is very important to the end users, is not to have to manually register. They want to know that the system is smart enough to know that they have just entered another local operator's territory. Equivalently, when we get into the notion of interoperability, we believe that without having a nationwide system or consortium of players who are using the same technology, when you leave one local operator and go to another, you may loose your features unless you are using the same technology standards and the same feature sets.

Finally, through our research we know that customers, when they call their customer service representatives to ask questions about their service, the last thing they want to do is hear a message to the effect, "well I'm not your local operator, let me transfer you to them so they can figure out what is going on with your account." MCI's vision is to have a centralized national service organization that would serve all the operators within the partnership.

In closing, I guess what I would like to say here is that MCI will pursue PCS and will probably pursue it three ways. The first way is in the SMR and the NEXTEL deal that was announced already. The second would be through the broadband auctions themselves. Again what we were thinking is that we would bid for those territories where we believe we would need to augment the NEXTEL spectrum that we currently have. So in those parts of the country where we don't believe that we need additional spectrum we just won't bid in those areas.

Finally, we are going to be a player as far as continuing to look for strategic partners. MCI does not think we need to be the local operator in every territory within the country, so we will look for strategic operators who would join in our joint venture and be the local operator for that territory.

26

The Importance of Unlicensed Spectrum

Jim Lovett
Apple Computer
Cupertino, California

I'll ask you to bear with me while I string together important issues that make unlicensed spectrum vital to PCS competitive issues, and PCS deployment issues in general.

The first time I came here to the MPRG several years ago, it felt very good to find fellow wireless advocates. Wireless, four or five years ago, was not an endeavor you were proud to be involved in. If you had a ham license, you tried to hide that fact in polite company, and denied it when you had a job interview, because it characterized you as being a bit "fringe-ish" in your thinking, and hardly professional.

Wireless has come quite a ways since that time. Something that has become evident is that the MPRG founders are certainly endowed with some of the most prescient thinking: insight and foresight and timing, that you can imagine. This timing has created graduates of the program who are going out and helping the industry develop; leaving here with an advantage over many of those who have been in the industry for a lot longer, because wireless, in the way we are thinking about it today, didn't exist until very recently. So, MPRG graduates are feeding industry right at a time when industry's need is greatest.

The second, almost scary part of MPRG's ability to time things, is that about two working hours from now, a big portcullis will fall across the doors of the FCC; the draw bridges will come up, and the staff and Commissioners will go into a rather monastic retreat and try to untangle all the noise and representations that have been made to them over the past months about PCS, especially since they released their Second Report and Order last October 22. That noteworthy event was immediately followed by some 80 Petitions for Reconsideration, including two from Apple, and I-don't-know-how-many lawsuits in Federal courts by parties objecting to Commission actions related to PCS. This session's timing is such that with the FCC's door closing today at 5 o'clock (not that anybody goes home at the Commission at 5 these days), this session may be the final input to the PCS process before what most people expect to be the definitive closure of discussions. I will try my best not to abuse the privilege of having Mr. Marchant here and listening; Mr. Marchant knows the computer industry's goals and objectives and we, in turn, feel our interests are being very well taken care of.

Having said that, I would like to give some background on how the unlicensed PCS process has evolved, and how I have been involved in it. Again, coming back to timing, this is a propitious celebration. Exactly one and one-half centuries ago last week, a noted American artist went to a table in a train station in Baltimore, flexed his fingers, and sent that now-famous query: "What Hath God Wrought?" That was May 24th, 1844.

Samuel Morse thoroughly intended to donate his invention to the federal government as an asset adding to the quality of life for all citizens. Of course, it didn't work out that way; the wired telecommunications network has been implemented by private interests, and it serves us well. It wasn't long, of course, until Morse's altruistic hopes were under-mined by independent businessmen seeking to broker technologies and networks and interconnects, which over the years developed into a need for regulatory structure and process to deal with conflicting needs, to sort out the winning technologies and advantageous ideas.

And so, a complex regulatory environment has slowly emerged to balance the powers. What we are talking about today, in terms of PCS values and competitiveness, will be resolved in good measure by the FCC next Thursday. There are billions of dollars and many jobs at stake; that's why this PCS activity has been so fervid.

If the title for this session had been slightly different, such as "The Economic Benefits of PCS," much of that economic benefit has already been gained because, on Apple's Petitions alone, the stack is about head-high of papers, pleadings, technical reports, and submissions written by lawyers. So one "economic benefit" of PCS has already been realized: it's one of the most effective "jobs programs" for lawyers we've ever experienced. (Fortunately they are good and very dedicated people.)

About the time I visited here about three years ago, I had just filed with the FCC a Petition for Rulemaking, asking for 40 MHz of unlicensed spectrum, for what I call "Data-PCS." I had several things in mind. Unlicensed operation was one of them, along with local-area coverage, high data rates as needed by computer users, no charges for the use of air-time, and the equipment itself to do its own technical management (in other words, to have the equivalent of what some people call "the FCC on a chip" built into the hardware).

I asked that the equipment authorization process would assure that equipment could live together, either interoperating or at least co-existing. The last point we asked for in the Data-PCS Petition was the opportunity for the industry to develop the operating rules, interactivity or coexistence rules, that were needed to assure some kind of reasonably sane operating environment.

We then formed an industry organization called "WINFORUM," which started out based in the computer industry, and we were fortunate enough to amass the support of virtually every major computer company and we experienced a friendly, collegial environment for working out some of those operating rules. At some point, the FCC realized, wisely, that this band could possibly be used for voice as well as data. So we invited the voice participants, particularly the wireless PBX people who were the most obvious voice participants at that time, to come help us define what we called the "spectrum etiquette," the coexistence algorithm.

The definition process then degenerated into one whereby the PBX people tired to keep the data people out of their way. That was not unexpected, because the last thing you want to do on a telephone system is lose a connection, or get a busy signal too often, or have a fragile, unpredictable service. For a telephone, one of the requirements is that once it is connected, it has to stay connected.

We data people are just the opposite: we look through the nooks and crannies and the time slots and the places to sneak through some more bits and pieces, and if we need it we will put massive amounts of overhead for error correction. We can do that; we don't have a timeliness dependency anything like the telephone user's time dependency.

After mighty labors, WINFORUM developed not a spectrum etiquette, but two different voice and data etiquettes. The FCC subsequently translated those etiquettes to separate voice and data bands. But to be more accurate, WINFORUM and the FCC made it even more complicated by producing three spectrum etiquettes: two for voice and one for data. The two voice etiquettes are, respectively, for relatively narrow band voice and for wide band voice. You can characterize these as being represented by CT2-like technology vs. CDMA-like technology (although that's an oversimplification, as anybody here would point out if I didn't'). So we currently have three spectrum etiquettes, not one, at least until a week from tomorrow. And right now, we have three frequency bands, and things are in a little bit of a turmoil.

The FCC also gave us a 40 MHz allocation for all these bands, which we appreciated after having seen it dwindle to 20 MHz. Although it re-grew back to 40 MHz, I'm apprehensive and also hopeful that the allocation will shrink again, a week from tomorrow, but it may be more practically usable, or perhaps some spectrum may be substituted that is not so burdened as the present allocation by incumbent microwave users.

The presence of microwave incumbents led to the formation of a group called UTAM, which started out as an ad hoc unlicensed wireless group but is in actuality comprised of eight PBX makers. UTAM is going through some very severe reality pains at this point. Only in Washington can you talk about $200 million or 800 million dollars to clear microwave stations. When you get to Cupertino, California, that looks like real money.

Computer devices have a special band-clearing requirement because they are characteristically "nomadic." They can be used anywhere, and they WILL be used anywhere. You buy them off the shelf, turn them on, and expect them to operate. This means we cannot deploy wireless products in a band where there are incumbent microwave stations that we could interfere with. We have, therefore, what we call the "last link" problem, meaning that the very last microwave-link station has to be moved out of the band. We have another "last link" problem: we have a hard time making people understand that computers are not telephones and do not follow the telephone metaphor; a highly controlled environment providing an extremely high guaranteed quality of service.

I'd like to turn now to what unlicensed PCS can be, and where it fits in, and I will wind my way to why unlicensed PCS is a competitive issue. Right now many of the open questions about unlicensed PCS center around data. Everyone wants to know "What's the killer application?" What's going to take unlicensed Data-PCS to success? There's no proof of market demand: nobody has implemented unlicensed wireless data devices yet, effectively and successfully in a commercial market, because there's no adequate legal provision for them: a classic chicken-and-egg quandary. The few companies that have launched wireless data devices have offered only wire-replacement LANs (for desktop computers), and they've generally found that nobody wants to buy them yet.

We think the market will change when you are able to walk around with your PowerBook or ThinkPad and connect wirelessly. Once we have nomadic, ubiquitous devices, we will need also to have someplace to go back to and connect with our fixed computers and wired networks. Trying to figure how the market will develop is like trying to predict what the traffic will be on a new bridge by counting how many people dive into the river and swim across it; until we have a demonstration example, a "bridge" that lets us communicate data at high speeds, we aren't going to know what Data-PCS will be used for.

I am going to take this opportunity to challenge in a friendly manner, Corey Eng's earlier presentation, about what he said regarding the capacity of ESMRs to provide services. I fully agree with what you said, Corey, and more power to you, but with one exception: you suggested that ESMRs can provide for ALL the wireless needs. What I think you described, however, were needs that are all generally characterized as "short messaging services." We at Apple, and the computer industry, are desperate for that functionality to become ubiquitous. But when you get to using your computer, you are talking megabits per second. You are talking about rich, multimedia applications. You are talking about school kids looking at volcanoes erupting, right then, in real time. You're talking about all kinds of social and medical and business opportunities that thrive on bandwidth.

That rich information environment leads to a significant competitive issue. It's a subject that you read about in just about every newspaper you pick up today. It's called the NII, the National Information Infrastructure. When the newspaper isn't talking about the latest Washington scandal that's attracting momentary attention, they revert to talking about the NII, and they, like the rest of us, are still struggling to define what the NII is and how it will be deployed. Earlier this year, Congressman Ed Markey polled the telephone and cable companies as to how they plan to offer schools some connection with the NII. They unanimously assured Mr. Markey they would provide NII "access," but their definition, to a single point at each school. In other words, they will provide a fiber or coax connection; through it they will pump their choice of material and ultimately they will charge a fee.

There was no mention of upstream services, or in-school networking. Last week in Washington I had the opportunity of asking a cable company executive if that was the extent of their plans. He answered, "Oh, no. We are going beyond that. We are going to provide connection for every classroom."

I said, "OK, keep going." He said, "What do you mean?" I replied, "How are you going to get to every student; how are you going to get to every desk; how are you going to get to every computer?" He said, "Oh, we'll do that too." How? He said, "We will run cables."

I asked, "Do you have any idea what that will cost?" He replied, "Well we didn't find it necessary to say anything about that because it is 'de minimus' problem." In other words, it is insignificant; buried in the underbrush, down in the noise level.

My wife looked up some numbers for me this morning. There are about 45 million students in 110,000 lower schools in this country, with more than two million classrooms. If you apply even a ridiculously low $100 per classroom to install cables, that's 200 million dollars. If you're talking real numbers you can see what that becomes. If you talk about asbestos being in some of those schools, you have no limit. (Our neighboring town of Mountain View decided to update their city offices so they could computerize. They figured it would cost them one-half million dollars to run some cables. Twenty-eight million dollars later, we have a new city hall because they took the first swat at the building and found asbestos; they had to level it and start over.) I don't think a few billion dollars is a 'de minimus' problem. Unlicensed wireless is part of the necessary solution for schools.

The American Indians offer another good pointer for the role of unlicensed wireless in the NII. We had a Community Networking conference two weeks ago at Apple, and a group of American Indians presented their involvement with the NII thusly: They said there were 500 federally recognized tribes and villages in the nation, but only one, in New York, has a significant connection to the Internet (one out of 500). There are 38 tribally controlled colleges; only one has an Internet connection. (There is no provision, today, for the Community wireless networks that could connect rural residents and communities to one another, to the NII, or to the Internet.)

How does that relate to competitive issues for wireless? There are two big issues that will be influenced by the role of unlicensed wireless in the NII, and as part of PCS.

First, all the charts, all the band-plans, all the technology roll-outs for wireless are based on winning an auction: buying airwave rights. I haven't heard anybody publicly address what will happen to auction-losers, but some will discuss this privately and a few have hinted at it in their filings to the FCC.

The scenario goes like this: there is to be an unlicensed band. It costs nothing to get the rights to use it. We don't have to buy the frequencies; we don't have to acquire licenses or combine licenses to get enough bandwidth. There is nothing stopping auction losers from using the unlicensed band to roll out PCS.

A case in point: the power utility companies. They have connection every few feet in virtually every building where people live or work, and a presence along every thoroughfare, far more access points than either cable or telephone companies have. There is nothing to prevent a power company from deploying unlicensed wireless devices that tie to that power infrastructure, and setting up a PCS. It need be no different from any other PCS except that it doesn't have such a massive startup burden.

Yes, unlicensed spectrum is unprotected spectrum, indeed. And at some point they (and we) will run out of it. But remember that we are talking about small cells, we are talking about 15 meter radius cells in the unlicensed band, not 8000 watts base stations that are serving whole communities as required for economic roll-out of licensed PCS. Frequency reuse factors can be monumental.

So we think that the next thing that is going to happen after the auction comes to pass is that people who didn't get the licensed spectrum are going to say, "OK, guess what?"

That opportunity has basic competitive issue ramifications. It provides a cap, for example, of what can be charged for licensed PCS service, especially in buildings and around campuses. You will find that people will deploy private systems such as wireless PBXs, wireless LANs, and maybe mixtures of the two, whose presence will put a ceiling on the revenue of what can be charged for a licensed service. The timing issue for PCS, that has been brought up by others, is also a competitive issue. Those who don't expect to get a license or don't think they can win one and are so precluded from activity in licensed PCS could start out very quickly to deploy unlicensed PCS, subject only to the finalizing of some rules for equipment compliance and measurement procedures.

There is yet another competitive characteristic for unlicensed PCS: the intensity of the information that is to be transmitted or carried by unlicensed PCS, particularly Data-PCS. Most of us use E-mail in our companies, we are using E-mail through the Internet to communicate with friends and relatives and colleagues. When we do this among ourselves, we don't expect to pay for that service. Our company will have an E-mail service set up (I suspect there is one here at VPI), that is essentially free. It is both transparent and free. Once you need to go into a wider area, you are going to need another kind of connection service, and that one may be tarriffed.

Tariffed and free services aren't in competition. One of the basic expectations for unlicensed service is that it will be a traffic generator for licensed services of all kinds, be they ESMR, cellular, CDPD, satellites, TelCos, wired systems, backbones of various types. Traffic generation will start in the local area; some percentage of that will filter into the wide area network. We think, again, that this is an extraordinarily important issue for PCS as it speaks for having both licensed and unlicensed activities working in concert.

I would like to finish up by saying that there is another activity in unlicensed wireless, that is the ETSI (European Technical Standards Institute) activity called HIPERLAN (High Performance Radio LAN). HIPERLAN is a 20 megabit per second functionality tied to an international standard now being developed by, among other, Apple. In Europe, there is an allocation for HIPERLAN at 5.2 GHz. There is no such allocation for HIPERLAN in the U.S.

One FCC gentleman asked me a few weeks ago, "What are you going to be coming back to us for, the next time you come to the spectrum trough? I replied, "An NII band." "What do you mean by that?" I replied: "Because you aren't giving us enough frequencies right now, we are going to go out and figure a way to get and use that "HIPERLAN Band," for HIPERLAN, of course, but also for other functionalities." We have in mind a wide range of technologies, geographic transmission extent, and bandwidths. You will therefore soon see yet another Apple spectrum initiative, that we hope others will help develop and implement, asking for the 5 GHz frequencies for an "NII Band" or "Internet Band" or a "Community Network Band."

In closing, I was told three years ago that asking for 40 MHz for Data-PCS was a little crazy. We did ask for it, and we got it. Asking for 150 MHz or 300 MHz for an NII band could appear obscenely greedy (or naive), until you read about Bill Gates' latest vision. He's asking for 4.4 GHz of bandwidth for his Teledesic network! That's a nice goal, we can't wait until it happens. In fact, why stop there? We're going to need more than that. Why don't we just get all the frequencies, and take them over to use them for our purposes, and pull the plug on the alternative: hundreds of MHz used to plug the sale of genuine simulated blue topaz earrings that, if you call right now, we'll send you the very last one at this incredible price, plus postage and handling. Have your credit card ready....

INDEX

A

absorbing half screen 131
adaptive array antenna 75
adaptive data rate 227, 229, 231-236
ALOHA 212
antenna diversity 197, 198, 200, 201, 203-205
antenna front-to-back ratio 162
ARPA 241
ARQ 211
AT&T 276
AutoCAD 185-187, 191, 193, 196
average bit error rate (BER) 79, 101
AWGN 108

B

BART 243
base antenna height 137
BEP 108
BER 79, 101
blind algorithms 33-35, 38, 41, 43-45
building 131
building penetration 20
Bulletin 10-F 157

C

CDMA (code division multiple access) 55- 72,
    107, 281, 287
CDPD (cellular digital packet data) 214, 227-229,
    231, 233, 234, 236, 237, 291
CELCO 281
cellular network 141
cellular radio 227, 229, 237
cellular telephony 64-66
CMA 76
CMOS 244
co-channel interference 43, 45, 46, 76, 102,167,
    228, 229, 231
codec 97
constraints and projection operators 123-125
continuity equation 132
cordless telephone 97
coverage 185, 186, 188, 190-196
CRC 211
CT2 287
CTIA 251, 253
CUG 209

D

Data-PCS 286
demodulation 97
differential equation 132
diffracted field 135
diffracted scalar field 134
diffraction 9, 131
diffraction parameter 135
diffusion 131
diffusion equation 133
digital FM 215
direct-conversion 97
discriminator 220
DMF 111
DoD 241, 242, 244
DSMA (digital sense multiple access) 227
DSO 208
DSP 263

E

earth radius factor 137
EIRP 157
equalizer 79
ERA 263
Ericsson 282
error correction 105
error variance 112
ESI 263
ESL 247, 248, 243, 263
ESMR 214, 272, 279, 288
ESN 214
ETSI 291

F

FCC 277, 279, 285
floor attenuation 12
FM click 216, 221
forward-difference method 135
forward error correction 70-71
frequency hopping 242
FRI 212

G

GAO 272
Gaussian approximation 60
GIS 180
GPS 179, 252
GSM 197-199, 201, 204, 205, 279

H

Hata model 137
Hazeltine 251
heat equation 133
hilly terrain 138
HIPERLAN 3, 291
host group addressing 207, 209
HP data 209

I

IFF 252
in-building propagation 10
indoor communication systems 8, 167
interference 197, 198
interference limited 4
interference rejection 31-33, 37-45, 47
interference suppression 33, 34, 38, 41, 42
interpolation 136
intersymbol interference 5
IS-95 59, 61, 64
ISM 2
ISO 252
isolation 158, 159
isolation hole 162
isotropic 136
ISTEA 264
IVHS 265

K

knife-edge 131

L

LATA 208
Lee Model 179-183
line source 132
loop antenna 144, 145

M

M-QAM 107
MACA 153, 154
macrocell 131
magnetic coupling 144, 145
MAN 209
MAP 107
Marquardt method 76
matched filter (MF) 110
MCI 273, 279
MDBS (mobile data base station) 227
media access protocol 163, 153

MF (matched filter) receiver 126
microcell propagation 19
MIP 264
mobile antenna height 137
mobile communications 76
Mobile Internet Protocols 244
mobility 197, 202
MOBITEX 207-214
modem 255
Motorola Microtec Communicators 242
MPAD 147
MTA 273
multipath 76, 98
multipath fading 227
multipath propagation 10, 57, 62
multiple diffracted waves 131
multiple knife-edge 135

N

nano-cellular network 144
narrowband PCS 216
near-far problem 61
near-field 141, 142
network control center 208, 214
neural networks 41, 43, 44
Neutrodyne Radio 251
Newton-Gauss method 78
NEXTEL 273, 279
NII (National Information Infrastructure) 289
NOKIA 282
Northern Telcom 282
Nyquist pulse shape 109

O

obstacle 138
Omnibus Budget Act 272
Omnibus Budget Bill 274
open area 139
optimum decision rule 111
OSI 208
outage probability 171, 172

P

parabolic differential equation 131
parasitic cellular 3
pass-loss 158
path loss 9, 131, 168, 186
path loss model 5, 11, 168, 189, 190, 192
PCMCIA 214
PCS 157, 241, 264, 277, 279, 285
PDA 1

PDI (Post Detection Integrator) receiver 126
PDP11 263
PICD 1
POCS (Projection Onto Covvex Sets) 122
point source 133
position location 66-68
posterior density 111
power control 61
propagation 10, 14, 158, 168, 169, 186
PSAM 108

R

R&D Program 242, 259
RAKE receiver 39, 63, 68-69, 125-126
Rayleigh fading 8, 182, 216
Rayleigh fading and log-normal shadowing 99
Rayleigh fading channel 110
RBOC 283
reciprocal 133
reflection 138
relative antenna height 137
Ricean fading 8
rms delay spread 14
road 138
ROSI protocol 207, 211, 212
RSS (Recombinant Spread Spectrum) 98
RSSE 115
RSSI 210

S

screen distance 137
screen separation 131
service reliability 101
SETA 255
shadowing 197, 202
$\pi/4$-shifted QPSK 78
simulcast 215
single knife-edge 133
SIRCIM 119
site specific propagation prediction 14
slow frequency hopping 197, 198, 200, 201, 203-205
small scale fading 8
SMR 214, 279
sniffer receivers 3
SP (Strongest Path Receiver) 126
spactial division multiple access 243
spatial processing 76
spectral efficiency 171
spread spectrum 30, 31, 41, 55-72, 97, 242
steady flow 132

steepest descent method 76
survey 31, 43

T

tapped delay line 76
TCT 108
TDD (Time Division Duplex) 100
TDMA 281
Telstar 251
temporal processing 76
terahertz optical fibers 243
terrain profile 132
time dispersion 12
time-vary MD 108
TQM 252, 253
traffic 197, 204
transceiver placement 185
transversal filter 75
Travinfo 248
tree 131
TRP 243, 245
T-R separation 160
TRW 243
TRW Phone Print 248
TRW Smart Search 248

U

urban area 137
UTAM 287

V

VDT 251
VISI-based DSP 264

W

WAMIS 244
whitening techniques 32-34, 40, 42
WINFORUM 286, 287
wireless LAN 185, 186, 196, 255
wireless PBX 277, 287, 290
W-PBX 1